Contemporary Television Series

Contemporary Television Series: Narrative Structures and Audience Perception

Edited by

Valentina Marinescu, Silvia Branea
and Bianca Mitu

CAMBRIDGE
SCHOLARS
PUBLISHING

Contemporary Television Series:
Narrative Structures and Audience Perception,
Edited by Valentina Marinescu, Silvia Branea and Bianca Mitu

This book first published 2014

Cambridge Scholars Publishing

12 Back Chapman Street, Newcastle upon Tyne, NE6 2XX, UK

British Library Cataloguing in Publication Data
A catalogue record for this book is available from the British Library

ISBN (10): 1-4438-5986-9, ISBN (13): 978-1-4438-5986-8

TABLE OF CONTENTS

LIST OF FIGURES

List of Tables

Chapter One

Watching the Cops: Police Perceptions of Media Representations of Police Work in British Television Crime Drama

Marianne P. Colbran

Introduction

For many people, media sources—newspapers, films, fictional and 'reality' television shows are the main source of knowledge and perceptions of the police (Skogan, 1990; Mawby, 2003; Huey, 2010; Reiner 2010). There is considerable evidence to bear this out. In the Policing for London survey, 80% of respondents said that the news media were their principal source of information about the police while 29% of respondents got their information from 'media fiction' (Fitzgerald et al. 2002). In the British Crime Survey (2006–7), 59% of people said they got their information from television and radio news, with 10% citing media fiction. As a consequence of this, as Huey (2010) has pointed out, there has been a growing body of work particularly in the United States on the question of whether fictionalized and 'reality-based' television crime programmes have any impact on audience knowledge and expectations of the criminal justice system (Stenross & Kleinman, 1989; Podlas, 2006; Tyler, 2006; Mann, 2006; Shelton, Barak & Young, 2007; Schweitzer & Saks, 2007; Coles & Dioso-Villa, 2007). However, the issue of whether the police themselves perceive that television portrayals of police work might have an impact on both public expectations of the police role and on public co-operation with the police remains an open empirical question (Huey, 2010). Some work has been done in a U.S. context (Arcuri, 1977; Perlmutter, 2000; Huey, 2010) and in an Eastern European context (Branea and Guguianu, 2013) but to date no work has been done in this area in the United Kingdom.

This study explores British police officers' perceptions of the influence of police procedurals on public expectations of the police and on public interaction with the police. Through focus groups with officers from two different forces—the Metropolitan Police Service (MPS) and the Greater Manchester Police Service (GMP), I look firstly at these officers' perceptions of mass media images of policing and how these affect police work and the public's treatment of the police. Secondly, I examine whether police officers from these focus groups perceive mass media images of policing to have an impact on public expectations of the police; and thirdly, I explore how police officers from the focus groups might wish the police to be represented in crime fiction in ways which, in their opinion, might increase public understanding and support.

Research

In this paper, I draw on the results of three focus groups—one from the Greater Manchester Police (GMP) and two from the Metropolitan Police Service (MPS)—in which participants were asked to watch an episode of *The Bill* and then to engage in a semi-structured group discussion. There were six GMP officers (two male, four female) and two civilian indexers in the first focus group; eleven officers (nine male, two female) participated in the first MPS focus group and in the second MPS group, four officers (all male) and two civilian indexers participated. The officers in the GMP group and first MPS group were all detectives—their ranks ranging in both groups from Detective Inspector to Detective Constable. In the second MPS focus group, all the officers were uniformed and their ranks were either Sergeant or Constable. In the study, I refer to respondents only by the police service for which they work and not by rank or occupational group in order to ensure anonymity.

The focus groups all started with a screening of an episode of *The Bill,* followed by a semi-structured group discussion. *The Bill* was a British police procedural television series that was broadcast on the ITV network from 16 October 1984 until 31 August 2010. The programme focused on the lives and work of one shift of police officers, both uniformed and detective officers, rather than on any particular aspect of police work. At the time of the series' conclusion, *The Bill* was the longest-running police procedural television series in the United Kingdom. I decided to show an episode of *The Bill* to these officers, partly because of its longevity and popularity but also because it depicted a variety of police work and roles. In addition, the show always placed particular emphasis on procedural accuracy and, for that reason, may have been seen to be particularly influential on audiences in terms of making claims for authenticity (Leishman and Mason, 2003). Participants in all three groups were asked questions about the episode they had just seen and questions related to other police procedural dramas. All questions were related to the core concerns of the study: (1) to what extent do officers perceive fictional representations of their work to have an impact on public understanding and interaction with the police and (2) what aspects of police work and the police role would they like to see portrayed in television drama in order to increase public understanding and decrease unrealistic expectations of police work.

Synopsis of episode shown to the focus group

***The Bill:* 'Corrupted', ITV 1**

Sergeant Stone and PC Sally Armstrong are called to a fight in a nightclub. Once there, Sally is pushed to the ground. One of the men involved in the fight goes back to help her. It transpires that he is an undercover officer from another station—PC Andrew Tipping—investigating drug dealing in the club. Tipping knows he will be arrested and asks Sally to take a stash of cocaine that he has on him, which he has just taken from a dealer. Sally agrees but regrets it. Meanwhile, the other person involved in the fight, David Bartlett, has been rushed to hospital. He has DC Jo Masters' card in his wallet. DC Masters is called to the hospital. Bartlett's daughter died at the club two years earlier after taking cocaine and her death was recorded as accidental. Bartlett has been doing his own investigation and believes that the undercover officer, Tipping, is involved in corruption. Meanwhile, Stone tries to pull strings for Sally to make sure no-one discovers she has covered up for Tipping.

Do police officers perceive that mass media images of policing have an impact on police work and interaction with the public?

The proliferation and expansion of mass media communications—radio, television, the newspaper industry and the Internet—has been one of the most significant social transformations of the last sixty years. One of the effects of this expansion of the mass media has been the social diffusion of images of the police.

Loader and Mulcahy (2003) have suggested that one of the consequences of what they describe as the 'routine visualization of policing on British television' (p.17) is that police dramas have enabled people to feel they know about the world of policing, and to fit their own encounters into this 'cultural imaginary'. This was echoed by one respondent from the GMP group, who suggested that:

'I think [police dramas] create the expectations of society, people's expectations of the criminal justice system, should they become witnesses, etc. The vast majority of people have very little contact, if any, with the police and they get their perception of it from the television. And when they do come into that situation, they react with that perception they've developed. If you see on a police show, that it's okay to say, I'm not going to talk to a police officer or to swear at them, then that's what they'll do. Television creates the perception for a lot of people because they have no real perception of dealing with the police.'

Officers from the Greater Manchester Police focus groups also believed that representations of police engendered expectations in the public as to how the police might treat them. For example, as one Greater Manchester Police officer pointed out:

> I think the thing for me is that it is so far from realistic, going up and talking to people or interviewing people. Everybody is aggressive with you or anti, everybody, as you said, carries on doing what they were doing. I mean, half the officers there, they never cautioned them or gave them the opportunity to have a solicitor present. It's just completely procedurally inaccurate.

Mastrofski (1999) argued that there are some fundamental expectations that should shape our treatment at the hands of the police, which include 'attentiveness, reliability, responsiveness, competence, manners and fairness'. Officers in the Greater Manchester Police commented on the high levels of aggression shown by both the public and police in their interaction on *The Bill* and contrasted it with their own experiences and need, as one officer put it, to get people onside and to trust them. However, although Foster (2003) argued that these principles tend not to be among the most valued of street policing culture, other researchers note a marked difference in the gendered quality of some street encounters. Braithwaite and Brewer (1998) found that 'the tactical choices of male officers more often placed them at risk of physical confrontation... Females were generally more supportive of citizens, preferring tactics which emphasized mutual power in the interaction' (p.286). Similarly, Loftus (2009), in her ethnographic study of two forces, suggested that in the main, a more service-oriented approach to street encounters was taken by female officers. As noted previously, the Greater Manchester Police focus group was made up predominantly of female officers and civilian staff and such perspectives may reflect a gendered reaction to interaction with the public.

When asked directly if they had ever found the public unwilling to engage with them, for example, because of something they had seen in a police drama such as *The Bill*, there was again a marked difference in response. Officers from the Metropolitan Police focus group said that none of them had experienced any problems with members of the public due to anything they might have seen on television, but officers from the Greater Manchester Police focus group indicated otherwise:

> If they go off what they've seen off the television and then they think, oh, I'm going to be a witness, I'm going to get petrol bombs through my front window, I'm going to get ran off the road, I'm going to get threatened and I'm going to get assaulted. Now I don't know about you but I know all the shootings we've worked on off Moss Side and I've never known one witness to get injured or actually ... but it's the perception it creates. And then when we come to investigate a job, people all say, "Oh no, I'd rather not get involved." And we say, "Why, what's been happening, what experience have you had that you'd rather not get involved?" and they say, "Oh no, I've seen it on *The Bill*, I'll be threatened, you'll give them my name and address, I'll have to go to court and stand up in front of others."

The Greater Manchester police officer suggested that, by portraying the handling of witnesses in a way that was unrealistic and out of date, the show might also create a fear of crime and a fear of the judicial system in the public.

However, both focus groups raised the point that, on occasion, television drama can be too realistic, especially in terms of depiction of forensic procedures. In the 1970s, respondents to Arcuri's (1977) survey on how police officers viewed fictional depictions of the work were highly critical of the way in which police dramas portrayed the resources available to police for their investigations. According to Arcuri (1977), one respondent observed that 'They, the public, feel that all experts are at our fingertips... as well as all up-to-date crime labs. In most cases, that's crap' (p.243). Writing over thirty years later, Huey (2010) elicited similar responses from respondents in her study of criminal investigators' perceptions of the impact of media representations of their work on public expectations. One of her respondents commented that 'CSI makes us look like idiots when we go to scenes, because they all think that they know exactly where a fingerprint can be found and what technique to do' (Huey, 2010, p.63).

Brunsdon (2000) argued that, since the late 1990s, there has been a move towards the medicalization of crime within the police drama genre, with programme makers increasingly depicting the police themselves employing new forensic methods to catch criminals, rather than this being the sole preserve of pathologists—a move which some officers find actively poses a threat to their work. One Metropolitan officer described how an episode of a police drama portraying criminals using caustic soda to cover their DNA had been used by offenders on a rape case in North London. He made the point that:

> The only thing I really object to on these sorts of things is when they do say how we deal with issues of forensic importance that are important to us and not commonly known outside of our fraternity. I mean, why on earth do you put all the cards on the table? We don't have that many cards left in our deck at all. So why do we tell the whole world how to do this, how to avoid that. If you're gonna show that sort of thing, could you at least send them down the wrong path.

In conclusion, officers from both forces raised concerns about the increasingly inaccurate and out of date portrayals of aspects of police procedure but, on occasion, officers conversely suggested representations of policing could be too authentic, particularly where forensic 'trade secrets' were being given away to potential offenders.

How do police officers perceive representations of policing might affect public perceptions and expectations of policing?

In his study of police attitudes towards media representations of policing, Perlmutter (2000) drew upon a study by Radelet and Carter (1994) to argue that citizens form expectations about the police from a variety of sources, including the mass media, and that if the police meet a member of the public's expectations, that person is more likely to have a favourable image of the police. However, Perlmutter (2000) argued that problems arise when those expectations are based upon unrealistic criteria and that one of the main sources of unrealistic criteria by which the public judge the police is the mass media and, in particular, television drama.

A number of commentators (Arcuri, 1977; Perlmutter, 2000; Reiner, 2003; Huey, 2010) have argued that, from the police's point of view, positive representations of policing can create almost as many problems for the police as negative representations, in terms of creating unrealistic expectations in the public about the police's speed and ingenuity of detection. This was a theme that was picked up by all the focus groups and they cited inaccurate timescales of police drama and how events were compressed and edited to render more rapid resolutions of the issue. Transitional but necessary events such as filing forms and completing paperwork are culled to make the narrative more interesting but, as one Metropolitan officer pointed out, this can have a negative effect on members of the public:

> I think that's the thing that gets me, totally unrealistic timescales. The problem is that people, whenever you speak to them, they say, oh, is it like *The Bill*? People have got this perception that our job is like that, and the fact that when we turn round and say it's gonna take a month to get this or it's gonna take weeks to get this, they think we're having 'em over, because they actually think we can get it straight away. And that's the problem, it's just totally unrealistic.

Officers were then asked questions about the storyline they had just watched and whether they thought specific details of the story might affect the public's perception of the police organization—i.e. the fact that this was an episode entitled '*Corruption*' and was about a young female uniformed officer's attempt to cover up for another corrupt officer.

Officers from all three focus groups believed that this episode had not shown the police organization in a favourable light and were concerned at the impact this might have on members of the public, particularly if asked to sit on a jury. One officer commented, 'They see that and they think, right, we're all, you know, corrupt.' Respondents from both focus groups commented that the storyline was inaccurate and showed a lack of research; that if an officer had been asked during the course of a raid in a nightclub to take a wrap of cocaine from another officer, the initial reaction would be to wonder if this might be an integrity test. However, the fact that the sergeant then attempted to cover up the young officer's mistake seemed to create the biggest controversy in the Metropolitan Police focus group:

> I thought the problem with it at the beginning was, the skipper, when she actually confessed, he had a go at her, and was saying, you know, outrageous, blah, blah, and then, he then goes on her side and then starts trying to cover it all up and there was a comment at the end, we all gotta look after our own and each other, and you're thinking well, you know, people do actually believe this is exactly true and this is what happens, and this makes 'em out to be corrupt, really...

The key line of the episode, which triggered the most controversy in the discussion, was at the end of the episode, where the young officer thanks the sergeant for covering up her mistake. The sergeant comments that, as police officers, 'we have to look after our own.' One Metropolitan officer commented that 'the general public do think the police all stick together, don't they? You are one big gang, one family and you have to look after each other and that's taken to the extreme. We know that's wrong.'

Although participants from both focus groups were at great pains to argue that the police had never been more accountable at any time in their history, and that any idea that the police 'looked after their own' was simply erroneous, many commentators have emphasized the internal solidarity of police officers (Clark, 1965; Westley, 1970; Cain, 1973; Reiner, 1978; Graef, 1989; Skolnick and Fyfe, 1993; Crank, 1998; Waddington, 1999; Foster, 2003; Loftus, 2009). Many studies also have stressed the codes of behaviour that enjoin officers to back each other up in the face of external investigations (Stoddard, 1968; Westley, 1970; Punch, 1985; Kleinig, 1996; Newburn, 1999; Loftus, 2009) though often, as Cain (1973) and Loftus (2009) point out, the offences that colleagues shield are not major but frequently attempts to conceal minor violations from supervisors. Thus, while internal solidarity and looking out for colleagues in beleaguered situations might be part of police culture, it would seem from the response of officers in both focus groups that this was an aspect of the 'backstage' (Waddington, 1999) of police work that officers would not wish to be made public on screen.

How might police officers wish to see themselves portrayed in television drama?

Participants from both focus groups, when asked how they would wish the police to be represented in television drama, argued that three important aspects of their work—black humour, the effects of some cases on police officers and what some officers called 'the reality of police work' or the service-oriented aspects of both frontline and detective work—were left out of current police drama. These were aspects that, in their opinion, would do more to increase public support for the police than attention to accuracy and detail.

Waddington (1999) has talked of how the police station canteen is a 'backstage area' where officers, whose actions are often invisible on the street, engage in displays before their colleagues and relate versions of events that affirm their world-view—that such banter is a palliative to the 'dirty work' (Hughes, 1962) of everyday policing. However, Perlmutter (2000) has argued, in his study of police officers' perceptions of media representations of their work, that the police officers he observed on patrol would see the inclusion of black humour in media representations as an unacceptable front-stage projection.

Garland and Bilby (2011) suggested that 'laughing and joking about the job, the villains and each other' (p.119) is a key ingredient of successful British dramas. However, officers in all three groups argued that the humour in the episode of *The Bill* they had just watched was too 'sanitized'. Instead, as one officer argued, the humour was simply not black enough and what was omitted from most police dramas was the way in which banter helps officers as a coping mechanism to deal with the horrors they see on a daily basis. One murder detective illustrated this:

> I'll give you an example. The team are all sitting there and the team all know they're going to watch a video of the murder scene. And you see people sitting there, laughing, joking, on ours, we've got this fella eating his second bacon butty and one of the girls says, "Bloody hell, you're gonna watch a murder scene." And he says, "If I didn't eat every time, I saw something gory, I'd be anorexic because that's all I do."

Another officer suggested that the depiction of humour in television drama might also serve to 'humanise' police work: As she suggested, 'if you get people to laugh with you, you've won half the battle to get them onside. And it makes us seem more human too, I think.'

Focus group participants also said that police dramas rarely portrayed accurately the effects of some cases on police officers. One Greater Manchester Police officer argued that too often the creators of television drama appeared to have very clichéd ideas of what stress did to police officers. As she commented, 'the way they show it at the minute in police drama is that every man goes home, drinks too much and beats his wife and every policewoman feels guilty about putting her job first.'

However, the same officer suggested that if the public were to see the effects on police officers of dealing with sudden deaths, cot deaths or dealing with victims, public support might be increased. She described the impact on a colleague who found the body of a missing child:

> He had searched it [the area where the child was found] and said he remembered seeing the bag but he didn't look in it, he thought it was rubbish and it absolutely destroyed him, absolutely destroyed him. She was already dead, he could never have saved her but the fact as he was saying, that her little body was left lying there two weeks-and you could see where rats had gone into the bag and things like that. Just haunted that man.

By contrast, researchers have suggested that, from very early on in their careers, officers of both sexes (Martin, 1979) learn that the expression of emotion is problematic. As Jackall (2007) suggested, 'the whole construction of the police world depends on officers maintaining the appearance of a rugged emotional distance, especially from the most emotion-laden and draining aspects of their work' (p. 236).

The last aspect of police work that both uniformed officers and detectives believed was missing from television drama was the service-oriented aspects of their jobs. As Foster (2003) has pointed out, despite the 'image of fast cars and catching criminals' (p.200), policing is frequently a more mundane and routine activity than is portrayed in the media, popular fiction and by police officers themselves. Many observational studies (Punch & Naylor, 1973; Skogan, 1990; Bowling & Foster, 2002; Loftus, 2009) have supported this and have suggested that, rather than speeding to the scenes of crime at high speed, in fact the police spend much of their time as 'the only 24 hour service agency available to respond to those in need' (Morgan and Newburn, 1997, p. 79). However, as Foster (2003) and Loftus (2009) have argued, the obsession with excitement and the emphasis in police working culture on crime-fighting being 'real' police work may lead to officers perceiving that some of the social service tasks they are often asked to perform are a waste of time and effort.

In his study of American patrol officers' perceptions of media images of their work, Perlmutter (2000) suggested that the officers he observed believed 'real' police work to be about crime-fighting. As part of his research, Perlmutter (2000) took photographs of the officers he observed at work and comments that when the officers looked at the photographs:

Overwhelmingly their preferences, the pictures that they stopped to comment on and ask for copies of, were action pictures: cops with guns raised, cops running, struggles with suspects, moments of anger on the street. Pictures displaying the procedural aspects of police work were flipped past without comment. (p. 13)

In the same vein, Perlmutter (2000) suggested that the officers he interviewed also believed such media images helped to shape public expectations of the police and that role strain was created by officers believing themselves to be under pressure to live up to that expectation. By contrast, the British uniformed officers and detectives in this study believed that more emphasis on the service aspects of both occupational roles would increase public understanding of policing and possibly even help to increase public support. Frontline officers argued for more storylines showing police work with communities while murder detectives argued for more screen time to be devoted in police drama to depicting detectives' care of victims and victims' families:

> I think Yorkshire were the first ones who said we've got to get a property bag that we can take property back. But a lot of people, they've lost a husband or they've lost a son and they can't open it. They put it away in a cupboard and they might decide two years later they can't stay in that house any more. So they go to move and then they find this bin bag and they don't even know what it is, they rip it open and bang, they're right back on the floor again. So you have to have something that is

> recognisable—how do you do it? Can't be black because that's rubbish bags. Let's have them green? No, green's for garden rubbish. The thought that goes into the bags...

As the same officer argued, such elements of care and thought on the part of the police for victims' families are rarely, if ever, shown in police dramas—yet in her opinion, such depictions could only increase support for the police and what she described as the 'reality' of police work.

As Loftus (2009) pointed out, there have been many developments in the British policing landscape over the last decade, many of which could be expected to transform police culture or cultures. There has been, according to Loftus, a notable increase in the number of minority ethnic, female, gay, and lesbian officers. Added to that, as Loftus (2009) has argued, another development in the police organization over the last decade, as a result of recommendations of the Macpherson Report (1999) which found the Metropolitan Police institutionally racist, has been 'the emergence of respect for diversity and recognition of cultural and gendered identities in policing discourse and practice' (p. x).

With that in mind, it could be argued that changes in the field in terms of a more diverse society and in terms of greater political sensitivity around policing may have led to changes in police cultures and that this is reflected in both uniformed officers' and detectives' emphasis on wishing to see the service elements of their role reflected in television drama.

Conclusion

In this study, I have attempted to explore whether or not police officers perceive that representations of their work in police procedurals such as *The Bill* have any impact on their interaction with the public and on public support for the organization. I have also attempted to explore what aspects of the police role or of policing in general officers would like to see represented in such dramas in order to increase public understanding of police work.

Through focus groups with uniformed officers, detectives and civilian staff from two British police forces—the Metropolitan Police Service and the Greater Manchester Police Service—I suggest that officers of both occupational groups do perceive representations of the police and policing in police procedurals to have an impact on interaction with the public. Officers gave specific instances of this—witnesses refusing to come forward in murder investigations as a direct result of representations of the treatment of witnesses on British police shows. Conversely officers argued that, on occasion, police procedurals could be too accurate in terms of depicting crimes such as rape or burglary and sometimes gave 'trade secrets' away to offenders, enabling them to elude arrest. Officers also expressed concern at the representation of their organization in the episode of *The Bill* shown to them as part of the overall discussion. Respondents argued that the sub-culture of internal solidarity among police officers gave a very negative and inaccurate picture of the organization. They also argued that the depiction was inaccurate as a result of extensive reform in the police organization, despite recent British police researchers' findings to the contrary (Loftus, 2009).

The second research question explored the issue of how police officers would wish to see their work depicted in television drama and which aspects, in their opinion, were never covered. Firstly, officers argued that the black humour they employed to help them deal with difficult cases was often 'sanitized' and they suggested that the depiction of such humour might serve to 'humanise' them in the eyes of the viewing public. This was in direct contrast to Perlmutter's (2000) study of American police officers' perceptions of media representations of their work, in which Perlmutter (2000) argued that this is an aspect of police culture officers would not wish to reveal to the public. Secondly, officers argued that too often television police drama depicts a clichéd view of the effects of stress on officers, showing both male and female officers as dysfunctional in their personal lives as a result of stress at work. Focus group respondents argued that depictions of the effects of certain cases on officers, particularly those involving children, might serve to increase public understanding of police work and the 'real' stresses of the job, even though this contradicts police ethnographers' findings (Martin, 1979; Jackall, 2007) that any display of emotion is seen as weakness in both male and female officers. Lastly, both uniformed and detective officers argued for a greater emphasis on the service-oriented aspects of the work of both occupational groups, despite previous ethnographers, particularly of patrol work, suggesting that this is the aspect of police work least valued by officers (Van Maanen, 1978; Foster, 2003; Loftus, 2009).

While it is only possible within the scope of the present study to suggest very tentatively that such findings may indicate a shift in British police attitudes and cultures, following the extensive reforms in the organization over the last ten years, it is certainly possible to suggest that police officers believe that media representations of policing can and do shape interaction with the public and, on occasion, impact on police work. It is also possible to suggest, given the influence of media representations on public understanding of the police role, that more portrayals of the service-oriented aspects of both frontline and investigative police work in crime drama might very well increase public understanding of police work and decrease unrealistic public and media expectations of the organization.

References

Arcuri, A. (1977). You can't take fingerprints off water: Police officers' views towards 'cop' television shows. *Human Relations,* 30, 237–247.

Banton, M. (1964). *The Policeman in The Community*. London: Tavistock.

Bowling, B. & Foster, J. (2002). Policing and The Police. In M. Maguire, R. Morgan & R.Reiner (Eds.), *The Oxford Handbook of Criminology* (3rd edn). (pp. 981-1033). Oxford: Oxford University Press.

Braithwaite, H. & Brewer, N. (1998). Differences in the conflict resolution tactics of male and female patrol officers. *International Journal of Police Science and Management,* 1(3), 276–287.

Branea, S., & Guguianu, A, (2013). *L'impact des séries TV policières sur les employés du ministère del'Intérieur en Roumanie,* paper presented at the Colloque La Police entre fiction et non-fiction, CNFS, 29 March 2013.

Brunsdon, C. (2000). The Structure of Anxiety: Recent British Television Crime Fiction. In E. Buscombe (Ed.), *British Television: A Reader* (pp.195-217). Oxford: Clarendon Press.

Cain, M. (1973). *Society and The Policeman's Role.* London: Routledge and Kegan Paul.

Clark, J.P. (1965). Isolation of the Police: A Comparison of the British and American Situation. *The Journal of Criminal Law, Criminology, and Police Science*, 56 (3), 307–319.

Cole, S.A. & Dioso-Villa, R. (2007). *CSI* and Its Effects: Media, Juries and the Burden of Proof. *New England Law Review*, 41, 435–470.

Crank, J P. (1998). *Understanding Police Culture.* Cincinatti: Anderson Publishing.

Doyle, A. (1998). "Cops": Television Policing As Policing Reality. In M. Fishman & G. Cavender (Eds.*), Entertaining Crime* (pp. 95-116). Bowling Green, Ohio: Bowling Green Popular Press.

Fitzgerald, M., Hough M., Joseph, I., & Qureshi, T. (2002). *Policing for London.* Cullompton: Willan.

Foster, J. (2003). Police Cultures. In T. Newburn (Ed.), *Handbook of Policing,* (pp. 196-227). Cullompton, Willan.

Garland, J. and Bilby, C. (2011). "What next, dwarves?": Images of Police Culture in Life on Mars. *Crime, Media, Culture,* 7(2), 115-132.

Graef, R. (1989). *Talking Blues: The Police in their Own Words.* London: Collins Harvill.

Hughes, E. C. (1962). Good people and dirty work. *Social Problems*, 10(1), 3-11.

Huey, L. (2010). "I've seen this on CSI": criminal investigators and the management of public expectations in the field. *Crime, Media, Culture*, 6(1), 49-68.

Jackall, R. (2007). A Detective's Lot: Contours of Morality and Emotion in Police Work. In S. Fineman (Ed.), *Emotion in Organization* (pp.227-40). London: Sage.
Kleinig, J. (1996). *The Ethics of Policing*. Cambridge: Cambridge University Press.
Leishman, F., & Mason, P. (2003). *Policing and The Media: Facts, fictions and factions*. Cullompton: Willan Publishing.
Loader, I. & Mulcahy, A. (2003). *Policing and The Condition of England: Memory, Politics and Culture.* Oxford: Oxford University Press
Loftus, B. (2009). *Police Culture in a Changing World.* Oxford: Oxford University Press.
Macpherson, Sir W. (1999). *Report of the Stephen Lawrence Inquiry.* London: HMSO.
Mann, M. (2006). The '*CSI* Effect': Better Jurors Through Television and Science? *Buffalo Public Interest Law Journal,* 24(1), 157–183.
Martin, S.E. (1979). Police Women and Police Women: Occupational Role Dilemmas and Choices of Female Officers. *Journal of Police Science and Administration*, (3), 314-323.
Mastrofski, S.D. (1998). *Policing for People.* Washington DC: Police Foundation.
Mawby, R.C. (2003). Completing the "Half-formed Picture"?: Media Images of policing. In P. Mason (Ed.), *Criminal Visions* (pp. 214-37). Cullompton: Devon.
Morgan, R., & Newburn, T. (1997). *The Future of Policing.* Oxford: Oxford University Press.
Newburn, T. (1999). *Understanding and Preventing Police Corruption: Lessons from the Literature.* London: Home Office Policing and Reducing Crime Unit.
Perlmutter, D. D. (2000). *Policing the Media: Street Cops and Public Perceptions of Law Enforcement*. Thousand Oaks/London/New Delhi: Sage Publications.
Podlas, K., (2006). The *CSI* Effect: Exposing the Media Myth. *Fordham Intellectual Property, Media and Entertainment Law Journal*, 16, 429–431.
Punch, M. (1985). *Conduct Unbecoming: The Social Construction of Police Deviance and Control*. London: Tavistock.
Punch, M., & Naylor, T. (1973). The Police: A Social Service. *New Society*, 24, 358–361.
Radelet, L. A., & Carter, D. L. (1994). *The police and the community,* New York: Macmillan
Reiner, R. (1978). *The Blue-Coated Worker*. Cambridge: Cambridge University Press
—. (2003). Policing and The Media. In T. Newburn (Ed.), *Handbook of Policing* (pp. 259-81). Cullompton, Willan.
—. (2010). *The Politics of the Police* (4th ed.). Oxford: Oxford University Press.
Schweitzer, N. J. and Saks, M. J. (2007). The *CSI* Effect: Popular Fiction About Forensic ScienceAffects the Public's Expectations About Real Forensic Science. *Jurimetrics*, 47, 357–364.
Shelton, D., Barak, G. & Young, S. K. (2007). A Study of Juror Expectations and Demands Concerning Scientific Evidence: Does the "CSI Effect" Exist?

Vanderbilt Journal of Entertainment and Technology Law, 9, 331–368.

Skogan, W. (1990). *The Police and Public in England and Wales: A British Crime Report Survey Report*, London: HMSO.

Skolnick, J. (1966). *Justice Without Trial*. New York: Wiley.

Skolnick, J., & Fyfe, J. (1993). *Above the Law: Police and the Excessive Use of Force*. New York: Free Press.

Stenross, B., & Kleinman, S. (1989). The Highs and Lows of Emotional Labor: Detectives' Encounters with Criminals and Victims. *Journal of Contemporary Ethnography*, 17(4), 435–452.

Stoddard, E.R. (1968). The Informal Code of Police Deviancy: A Group Approach to Bluecoat Crime. *Journal of Criminal Law, Criminology, and Police Science*, 59(2), 201-213

Tyler, T.R. (2006). Viewing *CSI* and the Threshold of Guilt: Managing Truth and Justice in Reality and Fiction. *The Yale Law Journal*, 115(5), 1050–1085.

Van Maanen, J. (1978). The asshole. In P.K. Manning and J.Van Maanen (Eds.), *Policing: a View from the Street* (pp.221-238). California: Goodyear.

Waddington, P.A.J. (1999) Police (Canteen) Sub-culture: An Appreciation. *British Journal of Criminology*, 39(2), 286–308.

Westley, W. (1970). *Violence and The Police*. Cambridge, MA: MIT Press.

CHAPTER TWO

CHILDREN'S PROGRAMMES AND THE NARRATION OF TV TECHNOLOGY

ÅSA PETTERSSON

In this chapter, I will discuss how TV technology has been narrated in several different ways over the years in Swedish public service children's programmes. The narratives in the children's programmes have been adapted to the TV institution's expectations regarding the child audience. Thus, studying TV narratives in children's programmes (cf. Lury, 2010, 2005) allows us to investigate how the TV institution imagines its child audience. It also allows us to examine notions of technology and a child TV audience as intertwined configurations (Woolgar, 1991) in the narratives of children's TV. I will provide examples of implicit and explicit use of technology in children's programmes and discuss examples in which the children on screen are more agents in their own right than followers of adult intentions with regard to TV technology, as well as examples in which the adult addressers are the ones introducing TV technology into the narration. In particular, this chapter deals with the fact that TV communication is one-way, but that, despite this, the narratives configure a joint 'we' between the TV institution and the imagined child audience (cf. Allen, 1992, Ang, 1991, Bignell, 2005, Buckingham, 2002, Edin, 2000, Hartley, 1987, Pettersson, 2013, Woolgar, 1991). This is accomplished both by ignoring the use of technology in TV production in the narratives as well as by using technology to tell stories in TV programmes for children. The analysis draws on TV Studies, Child Studies and Visual Studies to theorize and investigate how technology and children as a TV audience are co-produced, intertwined configurations in TV narratives (Corner, 1999, Ellis, 1993, Landström, 2009, Lury, 2010, 2005, Pettersson, 2013, Rose, 2001, Woolgar, 1991).

What stuck me when analyzing a large number of children's TV programmes is that the way in which programmes take up aspects of TV programme production has changed over time. This raises questions concerning what a 'proper' narrative for children is, what role TV technology is allowed to play in the narratives of children's programming and what this can tell us about the imagined child TV audience.

Children's public service programmes have existed since Swedish TV broadcasts began in 1956 (cf. Rydin, 2000). And since almost the very beginning, children's programmes have had a fixed time in the schedule. In this respect, children's programmes can be seen as a long-running TV series with a format recognizable over time and with hosts and characters who stay with the programme for long periods of time. Thus, children make up an audience that the public service TV companies have recognized from the beginning. Moreover, these companies are required to produce and broadcast programming for this group to fulfill their obligations to the state (cf. Pettersson, 2013). Here, I will present three examples of how TV technology has aligned with the TV–child audience configuration to produce narratives in children's programmes. These examples stem from research material from my PhD thesis *TV for Children—How Swedish Public Service Television Imagines a Child Audience*, which investigated TV programming for children from 1980–2007 (Pettersson, 2013).[1] Altogether the study investigated about 500 TV programmes for children, and the examples analyzed here are based on the analysis of the entire research material (Pettersson, 2013).

TV technology and children: what should children be told?

In the following, three different narratives from three time periods will be analyzed. The questions addressed are: In what ways are the technological aspects of TV production incorporated into the narratives for a child TV audience? And how are the technology and the child audience configured in relation to the TV institution?

Technology disrupting the narrative

In 1980, there was a public service monopoly on TV broadcasts in Sweden, and only two TV channels were broadcasting (cf. Hadenius et al., 2011). In one of them, the children's programme the *Half past four house* (*Halvfem huset*) was broadcast. It was a series that ran for several seasons, with 30-minute-long episodes that filled the whole children's programme

[1] A pilot study covering TV material from 2005–2007 was also conducted, and one of the examples discussed here stems from that study.

timeslot on Channel 1. The programme analyzed here contained several songs, an animated film, a saga about trolls and a band; moreover, both children and adults were present in the programme studio.[2]

The programme starts with an intro that resembles a filmed illustration in a children's picture book. When the intro ends, a man is seen walking around in what looks like a big living room and greeting a number of children, most of whom are sitting on a sofa in front of him. He introduces himself as James[3] and places himself on a stool in the middle of the room. He asks the children to look around the room; he says he thinks the room looks nice and suggests that a king might have lived there. Some of the children join him in talking about the story of the make-believe king and the room. The camera is moved to correspond to James and the children's talk, sweeping the parts of the room they are referring to. All of a sudden a child's voice is heard saying: 'So many cameras!' and another child's voice is heard joining this line of the conversation. This topic is, however, ignored by James, who instead focuses on his guitar; it is also ignored by the camera, which does not show any filming equipment at all. That the host and the production team are not interested in focusing on the technology or the TV production aspect of the setting also becomes clear a few seconds later. James is mostly filmed from the front. Between him and the camera, with their backs to the camera, are the children seated on a sofa and on the floor. James talks to the children sitting in front of him as well as to the audience by looking over the children and into the camera. The boy sitting closest to the camera on the right is looking quite attentively at James during the discussion of the room and he raises his hand to get attention. The other children are not raising their hands but talking freely, and the host does not recognize that there is someone waiting for his turn to speak. The boy's attention is, however, caught by the camera behind him, and soon he as well as some other children sitting close to the camera are looking straight into it and at the technological equipment near it. One boy picks his nose and stares into the camera and a girl also turns her head. This only takes place on screen for a second, as the camera focus is moved closer to the host, but when it moves away the boys comes into focus once again and one of them says: "Look we can see

[2] *Half past four, The half past four house* (Halvfem, Halvfemhuset) 1980, 19 September, Channel 1, 16.30-17.00. Produced by the Swedish Public Service Television Company.

[3] His full name is James Hollingworth, and he recorded several children's songs during the 1970s. Information on him and the programme can be found in the TV guide magazine for Wednesday 17th of September in 1980 (Röster i Radio TV 1980, no. 38).

ourselves on TV!" The camera focus then shifts quite quickly onto two other adults in another part of the studio, and the children looking into the camera are thereby moved out of the frame. The programme continues with an adult man reading a folk saga followed by a short animated film and live music. James and the children come back in focus as soon as the different programming parts have ended. But the camera is not placed behind the two boys again.

The programme addresses children using child drawings, animation, music and singing, but foremost by connecting the children in the programme and the child audience at home in its narrative. The host, James, tries to continue to build on this link by asking all children—both those in the studio and those in front of the TV—to look around the room, but the room that the host and the producers want to create for viewers is not sanctioned by all of the children in the studio.

When the children on screen focus on what they find interesting and talk about what they see, some of them show their opposition to following the narrative offered to them by the host. That this is not what the producers want can be seen when the camera moves away from the boys looking into it and focuses more closely on the host and other adults. However, the production thereby misses the opportunity to portray the children looking at the host and thereby to offer a position that the imagined audience can join.

The narrative—according to which the viewer is to join the audience on screen—is disrupted by the focus on the technological TV-producing equipment pointed out by the children on screen. It becomes obvious that there are things going on in the studio that are not shown on the screen and that the audience is by no means taking part in the on-screen activities there, but instead seeing only what the production team wants it to see. Thereby the talk about technology reveals that the viewers are not invited to join in on everything going on in this studio, but only to join the preferred narrative.

If the point had been to let the children in the studio act in whatever way they wanted, the technological aspects they pointed out would not have been ignored by the host or by the camera. This episode could be an example of when children are employed by adults to perform *children* in planned ways (cf. Lury, 2010), but where the children on screen are better characterized as agents in their own right than as followers of adult intentions. Nevertheless, it is the adults who are in charge of how the technology is used, and it is in their power to guide the narrative away from this topic.

This programme is built up using almost all the signs of doing child address, such as child drawings, picture-book images and picturing children (cf. Pettersson, 2013). Moreover, the adults—by trying to create a joint 'we' between the children on-screen and the child viewers off-screen—fill the programme with events that are linked to notions of child-related activities, such as sagas, songs and animations. The narrative is disrupted, however, when not all of the children act as though they are in on it, but instead some are more interested in the making of television and the technologies used in TV production. The technological aspect is not part of the preferred narrative, and hence it is not something the child audience should be aware of or take an interest in. However, this is something that would change over time.

Technology in the narrative

In 1992, the public service broadcasting monopoly had just been abolished and a commercial channel had entered the Swedish broadcasting arena (cf. Hadenius et al., 2011). The public service children's programmes still have a fixed timeslot, a set frame and a few hosts that alternate. The host Johan is always seated behind a desk talking strait into the camera when addressing the audience. He is not, however, only talking to the audience. To make the narrative of this programme work, he also talks to a rubber rabbit and to a never seen or heard technician named Bertil. When addressing Bertil, Johan looks out of the frame to his right, as opposed to facing the camera as he does when addressing the audience. Bertil is most often portrayed as causing Johan trouble, not wanting to change anything and not buying into Johan's ideas. Johan does not appear to think Bertil's job is particularly advanced, and in one episode he is heard saying: 'It is just a matter of pushing the button!' when Bertil hesitates to start a film.[4] But in regard to the technological aspects of TV programme making, Johan is also portrayed as dependent on Bertil to make sure that the short films, which are essential parts of this program, are put on and shown correctly.

On the desk in front of Johan a microphone is always visible. Johan also refers to the films by showing both film rolls and videocassettes. In this way, the TV technology is much more present in the narrative of this programme than it is in the example above. The fictive technician is needed to make the narration of the studio interesting, as Johan then has someone to talk to and argue with. Because Bertil also seems to be a bit

[4] *The Children's Programmes* (*Barnprogrammen*) 1992, 17 April, Channel 1, 18.15-18.45. Produced by the Swedish Public Service Television Company.

backward and stubborn, this creates a joint 'we' between Johan and the imagined child audience, as they are joined in opposing Bertil's somewhat boring attitude. The TV technology represented by the microphone, the film rolls, the videocassettes, and the technician also plays an important role in the programme narrative by making Johan and the studio meaningful and having a story to tell. Here it is made clear to the audience that TV is something created using technology: Even if Johan is speaking directly to the audience he needs a microphone in order to be heard and the technician needs to start the films in order for the viewers and Johan to see them. Nevertheless, this narration is also a fictive one, for example in that Bertil is never seen or heard. But the representation of how TV is made is also fictive, for example the sequences showing that Johan can just throw a film roll out of the frame and the film starts. In this program, technology provides the narrative with a plot as well as a way of allying the programme host with the imagined audience. There is no sign of hiding the TV production equipment from the viewers, as was the case in the previous example, but neither is there any attempt to explain how TV production really works. But, as we will see, there are ways of telling that story too.

Technology as the narrative

In 2005, the Swedish media landscape was largely digitalized and very diverse in comparison to the situation in 1980 as well as in 1992 (cf. Hadenius et al., 2011). Both the digital Children's Channel and Channel 1 were broadcasting public service children's programmes and the same host as in 1992, Johan, was one of the hosts of the flagship children's programme *Bolibompa*. This studio show framed all short programmes and films broadcast in the children's programme timeslot in 2005, and it still does. In the example analyzed here, Johan is going to film the control room of the TV studio.[5]

The whole narrative of the programme this day is built on showing and explaining to the audience how a TV programme is run behind the scenes. To accomplish this, a complex structure of technological gear is set up in the studio and all the way to the control room. A man named Björn is presented as the brains behind the set-up, and he hands Johan a camera with a long cord, which is connected to a screen placed in the programme studio. Johan then films his way to the control room and the viewers are invited to see this.

---

[5] *Bolibompa* (*Bolibompa*) 2005, 31 August, Channel 1, 18.00-19.00. Produced by the Swedish Public Service Television Company.

The whole sequence takes only a few minutes, and in this time Johan manages to get out of the studio, enter the control room and film the multitude of screens, buttons and lamps, which are what furnish that space. Two people, a man and a woman, are sitting in front of the screens. Johan asks the man if he is the one controlling all the buttons and the man admits that he is. The woman is introduced as the producer, and Johan says that she is in charge of making everything that is supposed to happen in the show happen, like starting the films, etc. She is then shown making the programme jingle appear on a screen and she says: 'But now you should be in the studio, Johan.' The woman is thereby also presented as Johan's boss, and Johan hurries to return to the studio. On his way back, just when he re-enters the studio, a cameraperson can be seen filming the studio setting.

In this program, the narrated plot is an investigation of how a TV programme is made and controlled, beyond what is usually covered by the camera. Johan and his technologically skilled friend Björn take pains to reveal what is actually hidden behind the scenes and to make these things visible for the audience. The audience watching the programme is invited to see this, and to see the people running the program. This differs from how the fictive technician Bertil was portrayed in the show from 1992; for example in 2005 there are no videocassettes being tossed out of the frame. Here the TV technology looks highly advanced and it is controlled by two people, who also tell the programme host what he should do. However, by showing the control room and the measures taken to make this possible, some things are also complicated. Instead of just presenting the cameraperson who films the studio everyday, Björn is invited to the show and the un-introduced cameraperson is to continue filming the studio, while Johan has a camera of his own. In this way, the narrative shows some of what is going on behind the TV screen, but it fails to question that the direct speech (cf. Allen, 1992; Edin, 2000; Pettersson, 2013) Johan directs to the camera is part of his conversation with the audience and not with the anonymous cameraperson who is omnipresent in the studio, but almost never seen or heard.

Conclusion

As we have seen, TV technology has been narrated in several different ways over the years in the Swedish public service children's programmes. In all of these examples, there is one common factor underlying how the producing institution has chosen to narrate and hence tell about the technology needed to make television. The ways in which the narrations have been presented have been dependent on how the programmes have imagined their child audience, and how that, in turn, has affected how the public service TV institution has hoped to maintain a relationship with this audience.

In the example from 1980, the production team wants a child audience that identifies with the children in the studio. Thus when the children in the studio do not focus on things that the camera is showing or that the host is talking about, this is not supported by the production. What becomes clear here is that there are children who can resist the adult power over the narration, but that they are not in control of the TV production technology. The child audience is thus imagined as being unaware of the technological TV equipment, and if it were made aware of the technology, the illusion of being one audience together with the children in the studio would be lost. Thus, in order to configure one joint audience for this program, the technology must be ignored in the narrative.

In the 1992 example, the TV technology is what makes up the narrative, as the fictive technician is a boring adult antagonist whom the host and the imagined audience are united in opposing. Here the fact that Johan cannot control the whole TV production himself is put forward as a self-evident truth, and the microphone and the different film artifacts are visual evidence of this. The fact that there is much more to TV production than can be handled by Johan and Bertil—even if he had been an actual person—is not discussed in this program, and there are no children in the studio who can demand real knowledge of the production process. On the contrary, what is needed to make the conversation and relation between the host and the imagined child audience as 'real' and as intimate as possible is that there is someone who is uninterested in forming a closer relationship with Johan as well as with the audience. In this way, the audience is informed about the production aspect of TV, but the way in which this is narrated causes the direct address that Johan uses to form a closer relationship with the audience. This narration configures a tight bond between Johan and the imagined child audience on the basis of the shown technology—thus quite the contrary to the former example.

In the 2005 example, Johan wants to inform the imagined child audience about what is going on behind the scenes, and he does so by using technology to show other kinds of technology. In this program, he talks to the audience directly, speaking into the camera (cf. Allen, 1992, Edin, 2000: Pettersson, 2013), just like in the other examples, but he also shows the audience what is really happening in the control room: who starts the films and who is really in charge of the programme broadcasts. However, because the cameraperson in the studio is not introduced, one mystery of TV production remains here too. In children's TV programming, the narration of a possible two-way-communication between the host and the audience is never revealed as an illusion.

The imagined child audience is spoken to directly from the screen, and even if the TV technology used to produced television is gradually allowed into the narrative in children's programmes, the fact that TV is a one-way communication medium only is not something that can be spelled out in the narrative of children's TV programming, and this has continued to be the case over the years under study (cf. Pettersson, 2013). On the contrary, the amount of technology allowed into the narration is limited so as to maintain the configuration of a joint 'we' in the communication between the children's programmes and the imagined child audience of public service TV. The intertwined configurations of technology and the child audience are thereby what keep the illusion of the child TV narrative going. As we have seen, this can change but not in a way that would call into question the closeness of the public service TV institution and its child audience.

References

Allen, R. C. (1992). Audience-Oriented Criticism and Television. In R. C. Allen (ed.). *Channels of discourse, reassembled: television and contemporary criticism*. London: Routledge.

Ang, I. (1991). *Desperately seeking the audience*. London: Routledge.

Bignell, J. (2005). Familiar aliens: *Teletubbies* and postmodern childhood. Screen, 43 (3), 373-387.

Bolibompa (*Bolibompa*) 2005, 31 August, SVT, Channel 1, 18.00-19.00.

Buckingham, D. (2002). The child and the screen. In D. Buckingham (ed.) (2002), *Small screens: television for children*. New York: Continuum.

The Children's Programmes (*Barnprogrammen*) 1992, 17 April, SVT, Channel 1, 18.15-18.45.

Corner, J. (1999). *Critical ideas in television studies*. Oxford: Clarendon.

Edin, A. (2000). Den föreställda publiken—programpolitik, publikbilder och tilltalsformer i svensk public service-television. Diss. Stockholm: Symposion.

Ellis, J. (1993). *Visible fictions: cinema, television, video*. Rev. ed. London: Routledge.

Hadenius, S., Weibull, L. & Wadbring, I. (2011). *Massmedier: press, radio och tv i den digitala tidsåldern*. Stockholm: Ekerlid.

Half past four, The half past four house (Halvfem, Halvfemhuset) 1980, 17 September, SVT, Channel 1, 16.30-17.00.

Hartley, J. (1987). "Invisible fictions: Television, Audiences, Paedocracy, Pleasure". *Textual Practices*, 1 (2), 121-138.

Landström, C. (2009). En queer blick på teknik i TV-serier. In *Tidskrift för genusvetenskap*, (2-3), 69–91.

Lury, K. (2005). *Interpreting television*. London: Hodder Arnold.
—. (2010). *The child in film: tears, fears and fairytales*. London: I. B. Tauris.
Pettersson, Å. (2013). *TV for children: How Swedish Public Service Television imagines a child audience*. Diss. Linköping: Linköpings universitet.
Rose, G. (2001). Visual methodologies: an introduction to the interpretation of visual materials. London: Sage.
Rydin, I. (2000). Barnens röster. Program för barn i Sveriges radio och television 1925-1999. Stockholm: Stiftelsen Etermedierna i Sverige.
Röster i Radio TV 1980, nr. 38.
Woolgar, S. (1991). Configuring the user: the case of usability trials. In J. Law (ed.), *A sociology of monsters: essays on power, technology and domination*. London: Routledge.

CHAPTER THREE

QUEBEC TELEVISUAL FICTION AFTER REAL TV: MORE COMPLEX AND MORE ARTISTIC TV SERIES

YVES PICARD AND PIERRE BARRETTE

Perhaps the 21st century will prove to be the golden age of TV, a time in which the handsome possibilities of high-definition digital programming allow television to become an art form every bit as adventurous as cinema (Newman and Levine, 2011, p. 100)

TVseries have been the subject of sustained interest amongst scholars since the beginning of the new century. To mention some of the work carried out during the first two years of the present decade alone, we have seen that one scholar (Butler, 2010) is studying style in American televisual fiction, while another in France (Esquenazi, 2010) asks whether TV series represent the future of cinema. The following year, another French scholar (Pichard 2011) documented what he describes as a new golden age of television series, while elsewhere in the world (Perez-Gomez ed. 2011) scholars reflect on what is, for them, television's third golden age. That same year, two American scholars (Newman and Levin 2011) took the measure of the scholarly community's newfound interest in television: the medium is in the process of being legitimated.

This change of perception derives from a transformation of the products on offer. New televised fiction, often described as 'quality television' (McCabe and Akass 2007), is taking advantage of high-definition digital technology and increasingly larger screens, whose appearance coincided with the arrival of the new century. These quality television series, contrary to commercial cinema, which seems increasingly turning its sole interest towards comic-book heroes and romantic comedies aimed at a teenage audience, often offers complex storylines and controversial themes aimed at a more adult audience.

The goal of the present article is to continue in this line. On the one hand, the body of work we will discuss falls into this period: we will examine television fiction in Quebec, which according to our research (Barrette 2006; Barrette 2008; Picard [2009] 2010; Picard 2011; Picard 2013) has been contributing to the quality phenomenon since 2004. The Quebec TV corpus is little discussed in scholarly research, yet it remains highly popular despite the digital tsunami, as the weekly ratings published by the firm BBM demonstrate. To give just one example, but which has the force of symbol, note that if the event that is the *Superbowl* reaches over 110 million viewers, a third of the U.S. population, one Sunday a year, viewing habits in Quebec are such that the two most popular programmes have more than 3 million people, or almost a half of the Francophone population of 6.4 million, *every* Sunday night. On the other hand, our thesis is that quality television fiction is also a response to reality shows, which, in the case of Quebec, appeared on the landscape a year earlier (*Loft Story*: TQS 2003-; *Star Académie*, TVA 2003-). More precisely, we will focus on two Quebec fictional television series, which reference reality shows, with an emphasis on the first and last episodes of each of these series. These two series are available on DVD (region 1).

The first, *Les Invincibles*, is written (in collaboration with François Létourneau) and directed by Jean-François Rivard and was broadcast on the public television network, the Société Radio-Canada, from 2006 to 2009. It numbers thirty-five episodes, each forty-three minutes in length, over three seasons and tells the story of four thirty-something men who decide to sign a pact: to prolong their youth, they will all simultaneously leave their girlfriends. Alongside this main storyline, short scenes show the four heroes speaking candidly to a fake documentary film crew, while comic-book vignettes also appear from time to time. The second, *Tout sur moi*, is written by Stéphane Bourguignon and directed by Stéphane Lapointe and was also broadcast by the Société Radio-Canada, in this case from 2006 to 2011. Its five seasons number sixty-five episodes of twenty-six minutes each. A critical and popular success, often with a wacky humour, it features the professional and romantic tribulations of three actors who have been inseparable friends since theatre school and play their own roles in the show. By his own admission, the writer Stéphane Bourguignon drew on the strong friendship between his girlfriend and the two other actors to write a fictional story based on real characters and true situations in order to parody and pervert them.

Les Invincibles... at the beginning

The pre-credit sequence of the first episode of *Les Invincibles* is disconcerting: it takes the form of a montage of previous episodes. Right from the first image, before the story even begins, a screen title makes the unexpected announcement: 'Previously on *Les Invincibles...*' It gets better: three of the four male characters are shown speaking about the urgent need to enter the fictional story, referring to the 'first episode, which begins tonight,' while the fourth declares himself reticent, announcing that he will 'lose everything' if he agrees to appear in the show. To add to the confusion, the sequence includes the image of a clapperboard from the shooting of the show. One might suspect that this technique could give rise to two different reactions from the audience. The first is rejection: how can one summarize what has not taken place, especially given that one character appears to resist participating? The second possible reaction, on the other hand, is to be drawn in all the more powerfully, given that it is based on the sophisticated complicity which accompanies the meta-textual: it involves acknowledging the convention of television fiction that is divided into acts—the pre-credit summary—and enjoying the *transtextual* (Genette1997) variations on offer. At the same time, it admits the inserted expressive device—*the fiction within the fiction*—and interprets it as a 'reflexive construction' (Metz 1991) subject to the arbitrariness of the enunciation. This opening may have thus disappointed a good many viewers. Indeed, in François Récanati's (1979) words, it is far from transparent, but rather resolutely opaque, demanding that the audience examines its own relation with the work in order to make sense of it. At the same time, it has everything required to make it possible for other viewers to be drawn in—those experienced in viewing modes less associated with popular culture than with modern cinema. The opening scene in *Les Invincibles* plainly offers up an expressive complexity which makes considerable demands on the audience, requiring them to appreciate the work's *metalepsis* (Genette 2004).

The pre-credit sequence does not stop there. It includes a second part, separated from the first by a fade to black, which begins with a superimposition: 'Now...". This scene is set in the present. The four young men are seated in a basement around a table like card players, entering into the pact of the storyline: they will leave their girlfriends to maintain their freedom and will indicate that the deed has been done by wearing especially child-like blue wristwatches. The sequence does not seem out of the ordinary, but it turns out to be reflexive twice over. On the one hand, it underscores the fact that the *enunciative* contract established with the audience during the previous sequence is accompanied by another, of a *deceptive* nature: the four main characters are carrying out an immoral act, that of leaving their girlfriends for the stated purpose of regressing. The audience is thus invited to react either by rejecting the premise—if they base themselves on logic—or by accepting their complicity from the outset. The gesture may be deemed pointless by some and perceived by others as announcing a programme full of existential and narrative digressions. At the same time, the sequence also appears to be reflexive by virtue of its dialogue. The recalcitrant character seen in the previous sequence laments the storyline's predestination in clear terms. After a pause, he concludes: 'I have a feeling this will end badly.' These words are charged with meaning, whose full extent we will only come to know when the story is complete, and are immediately followed by the credit sequence's rock music, giving them greater weight. This allusive reflexivity involves a great deal of complicity with the audience: we will only know three years later if the character is right. We can also see all this differently: because audiences may now view the show's episodes intermittently, one after the other, through DVD, VOD or on the Internet, its creators conceived this Quebec TV series as a long narrative developed over time. Quality television is also based on new television viewing habits.

The credit sequence of the first episode of *Les Invincibles* appears to contradict the scholarly viewpoint expressed at the beginning of this article. The series' actors and crew are introduced by means of a sequence of brief, colourful depictions in a comic-book vein set to rock music, which seems to proclaims the show's commercial appeal. The credit sequence appears to aim to please an audience that is naturally resistant to the complexity and opacity of the Quebec TV series. This is an illusion. On the one hand, the mimicking of commercial cinema's codes constitutes a shrewd *transtextuality*: the TV show, aware of its media competitor, the cinema of comic super heroes, incorporates it in a fashion worthy of its title, *The Invincibles (Les Invincibles)*. This *transtextuality* is also cynical: the four heroes, as we have mentioned, display behaviour contrary to that found in heroic fantasy; they claim the right to immaturity. In this sense, they are invincible only in their own minds. This mimicry is thus reflexive and may be read in addition as a critique of the artificiality of Hollywood action heroes. On the other hand, the comic-book graphics of the credit sequence are not mere affectation: they will prompt three reflexive references. First, the comic book is taken up in the storyline, because the recalcitrant hero dreams of the day he can be a comic book artist. Second, they recur in the show's discourse, because sequences depicting the four heroes alongside their super-hero comic-book avatars are inserted into the narrative as discursive commentaries on the *diegesis*. Finally, they are incorporated into the *diegesis* like the material expression of the imaginary world of the recalcitrant hero: elements of it appear on screen throughout the series, in a denial of any sense of verisimilitude. The credit sequence of *Les Invincibles* thus constitutes a doorway onto multiple and opportune self-referencing. Quality television engages viewers in a protean reflection and demands that their attitude towards the work shift with the changing scenes. Indeed, new TV complexity engages the audience attitudes.

The next sequence depicts discussions between the four heroes and their respective girlfriends, underscoring the similarity of the romantic relationships in which the four men are involved, as well as the virtuosity of the scriptwriting and direction: in the repartee, the characters respond to one another with fluidity even though they are speaking in parallel spaces. That said, what comes next could lead one to think that expressive complexity has given way to narrative simplicity, albeit well-presented, were it not for the successive appearance after the first commercial break of two startling sequences for a show such as this, signalled by a graphic *I* not unlike that found in *The Incredibles* (2004). In each sequence, one of the four characters discusses his situation with a film crew. In the first, the initiator of the pact describes his satisfaction with his decision, while in the second the recalcitrant hero describes his work with feigned nonchalance. In each case, the character is looking off-screen, and the mise en scene distances the viewer, thereby mimicking a convention of reality television: the confession (Jost 2009). Although reality shows use the technique to describe the character in a manner consistent with the role for which they were cast, and as a result to comfort them, in *Les Invincibles* it gives rise to the opposite effect, thus generating a critical distance: the protagonists betray their motivations in ways which elicit discomfort. The sequence has a meta-textual quality, like the comic-book sequences, which is to say like a layer of meaning, which reflects the *diegesis*: it offers a counterpoint. Networks of meaning thus collide: the story of the four men is reflected concurrently on the one hand by a comic book which shows it to us in an ideal form of the self and on the other hand by a confession in the form of an acknowledgement of the self. Clearly, the linear must be seen in a tabular manner to fully make sense. It is thus not surprising, that *Les Invincibles* was scorned as an apologia for immaturity, even by scholarly criticism (Létourneau 2010), when it seeks to distance itself from it. Quality television demands a change in viewing postures on the part of all audiences, both popular and scholarly.

Les Invincibles... at the end

Thirty-four episodes later the show came to an end, concluding three seasons structured like a trilogy, which, as we have seen, was conceived from the beginning as an organic whole. The show's predestination was announced twice: by the recalcitrant hero's stated unwillingness, right from the false summary of previous events, to enter the fiction; and, after signing the pact, his remark to the effect that it would all turn out badly. We may be tempted to ask ourselves, however, who was most concerned by this future threat hanging over the story. The pre-credit sequence of the final episode suggests the answer. Instead of summing up, it presents a flashback, set in 1983, identified as such by cinematic codes (widescreen image, a slow, smooth tracking shot, lighting for effect), which would turn out to be as much *trans-textual* as reflexive. The then wife of the recalcitrant hero, a child at the time, already reveals her brittle nature when discussing the care a newborn should be given so as not to affect its future. This opening lets us think that, by means of a tragic reversal, she herself will perhaps be affected by this maxim, the way scripted films often have such things in store for their characters. It also demonstrated that the TV fiction takes today the form of a *cinematization* (Buxton 2010; Newman and Levine 2011), which demands in return from the *lector in fabula*, to employ Umberto Eco's expression, that they adapt their reading practices to the medium's evolution. In the same way that art-house cinema gave rise to a wave of film studies oriented towards an abundance of academic studies with the development of semiology under the impetus of Metz (1990), quality television has brought television studies to consolidate its theoretical models and led audiences to change their perceptions.

The next sequence connects the past with the present. It shows the wife, in the present, in the midst of an emergency: she appears to be about to lose her child during childbirth. This sequence would be a transitional scene without interest if it weren't for the fact that the warning hammered home in the flashback seems already to be turning against the person who uttered it, and that we are in the presence of a reflexive construction. A friend of the frightened woman casts a glance off-screen in the direction of the bed, but the cut-away, which should validate the situation, is concealed. In an effect, which must be described as an *auteur*'s intrusion, the expected counter-shot gives way to a fade to black. The following sequence takes the reverse course: it gives the answer without asking the question. Alongside the superimposed opening credit sequence, the images show the doorway to the home of the recalcitrant hero, unaware of the tragedy unfolding and who, for this reason, does not look in the direction of the bed, while the camera, in a noticeable display of autonomous gesture, shows a bloodstain on the sheets. By denying us the expected image and by offering us what remains unknown to the protagonist, the mise-en-scene is doubly arbitrary in its interventions and the subjectivity of its presence. Far from being anonymous, as it is with what John Thorton Caldwell (1995) and Jeremy Butler (2010) call the *zero-degree style*, the direction adopts what must be called a *second-degree style*, as we have described elsewhere (Barrette and Picard 2013): a style which is both original and personal, or in a single word *authorial*. As such, it sets new standards for the viewer, requiring that the television object be appreciated for the degree of expression to which it departs from the aesthetic norm in a meta-critical perspective. A reflexive construction, in Christian Metz's (1991) sense of the term, involves a *transtextual* reading, not in order to make visible the convergences between a text and other texts but to make visible, rather, the divergences between a text and other texts. That said, not every audience knows how, or is able, or even wants to do the task. Quality television is continually challenging the *viewer in the text*: it requires of him/her an ongoing proactive attitude.

The finale of the last episode of *Les Invincibles* takes the form of an anthology[1]. The narrative arcs of each of the four protagonists come to a head in a series of sequences connected by non-diegetic music, the song *I'm Still Loving You* by the Scorpions, which runs for the final six minutes. The events shown have a tragic tone. The recalcitrant hero finds himself alone with the pain resulting from his companion's hospitalization and losing battle with death. His chagrin, which is intensified by the comic-book images cut into the editing, is made poignant by the fact that the other three members of the group turn their backs on him, as he had demanded and gives rise to an incongruous materialization of his imaginary world. The threads that the finale of the show ties together are not only complex and reflexive, they are *transtextual* and they strengthen the predications formulated at the outset of the narrative: the recalcitrant hero loses out in the end; they echo one another and describe characters who are not Manichean like the comic-book heroes they identify with; and they stand out from standard television fiction by setting aside anecdote in favour of distanced perspective, encouraging the use of crosscutting set to music whose lyrics function as a commentary. In this sense, the *transtextual* link to the finale of the first film in Francis Ford Coppola's *Godfather* trilogy (1972), in which crosscutting depicts events

[1] The segment is available on YouTube at:
http://www.youtube.com/watch?v=uuBWzAhjNbg

simultaneous with the hero's decision, adds another layer of meaning to the finale of *Les Invincibles*. It is a forceful example of how this Quebec TV series is conceived in the second degree style and contributes to the *cinematizing* of the small screen, as David Buxton and Michael Z. Newman and Elana Levine argue. That said, we cannot forget that the complexity of *Les Invincibles* demands an open-minded audience, capable of working with the show to unlock its meaning. Otherwise, it is without effect.

The climax of *Les Invincibles* is exemplary in another respect. Because the characters are locked inside their respective solitudes, it consists of a series of images in which words become increasingly obsolete. The expressiveness of the finale of this last episode of this fictional twenty-first-century Quebec television programme thus reveals an essential property of quality television. As John Thorton Caldwell and Jeremy Butler have remarked, speech is the structuring element of zero-degree style to the point that in soap operas, as Butler remarks (2010, 53), the 'images are constructed to illustrate the words, rather than vice-versa.' In this rudimentary aesthetic,the visuals are at the service of spoken words; speech thus determines the directing, composition and rhythm of televised fiction. The finale of *Les Invincibles*, however, proceeds in the opposite way: during the last six minutes, the images appear gradually to be released from their duty to record the talking heads and signify on their own. The production of meaning thus reflects the presence of a different king of author: the voice of the scriptwriter is silenced in favour of the vision of the director. To borrow from the field of cinema studies, *monstration*, as André Gaudreault (2009) has identified it, yields centre stage to *enunciation*, in Christian Metz's (1991) sense of the term: the *attractional* of the voice is replaced by the *authorial* of the gaze. *Les Invincibles* finishes its complex textual course not on talking heads, but, with the aid of the Scorpions, instead on silent pictures in full *(e)motion*.

Tout sur moi... in the beginning

Airing a year before *Les Invincibles* and coming to a conclusion two years after it, and leaving the sixty-minute existential drama aside in favour of the thirty-minute situation comedy, the pentalogy *Tout sur moi* took up the baton from the trilogy *Les Invincibles* and set the bar of twenty-first-century Quebec TV fiction even higher: it is more complicated than the latter, more openly artistic, and also more demanding. *Tout sur moi* maliciously blurs the lines between reality and fiction: in each episode, three young actors, whose first names and professional careers are the same as the real actors playing them, experience, as the advertisements claim, their share of 'romantic ups and downs, obsessive craziness and professional humiliation' (Radio-Canada.ca). The result is a black-humour TV fiction in which the three heroes are constantly getting themselves into uncomfortable situations in which their difficulties compound, all the more so in that the boundaries between true and false are blurred. To add to the complexity, other actors and Quebec cultural personalities join in the fun and play themselves in scenes in which it's difficult to tell if the scene is real or imaginary. This Quebec TV series, with its telescoping of levels of reality, must be resolutely seen indeed in the second degree to make sense. Like *Les Invincibles* but with even more audacity, and undoubtedly to greater effect, it pokes fun at reality shows, which claim to depict reality, allowing the viewer to grasp instead that every representation is a simulacrum. While TV reality show producers appear to deny that they have read Jean Baudrillard (1984), the producers of *Tout sur moi*, for their part, do not deny it: the title of this Quebec TV series ('All About Me' in English) is a nod to the fiction film *All About Eve* (Joseph L. Mankiewicz 1950), which is also structured around actors, the acting profession and false appearances. This title, with its hints of Guy Debord's (1967) society of the spectacle, Christopher Lasch's (1979) theories of narcissism or Gilles Lipovetsky's (1983) notions of contemporary emptiness and individualism, also thumbs its nose at reality TV: it asserts without equivocation that fictional depiction is a simulacrum.

Tout sur moi references reality shows in order to distance itself from them. At the same time, we could say that *Tout sur moi* is one more variant of the situation comedy to take us backstage in the entertainment industry, like *The Larry Sanders Show* (HBO 1992-1998) or *Entourage* (HBO 2004-2011), or to feature uncomfortable situations in this milieu, such as *Curb Your Enthusiasm* (HBO 2000-). But there is also the form. That of *Tout sur moi* does not take up where a mockumentary such as *The Office* (BBC 2001-2003; NBC 2005-2013) leaves off, with its simulations of a film crew and the of reactions both on camera and backstage during confessional scenes. It presents itself, rather, as a fiction,whose floating boundaries between true and false contribute to sequences depicting singular confessions, presented as theatrical soliloquies. The three pals, situated in a timeless, brightly coloured space, comment on their situations and break down the fourth wall by looking at the TV viewer 'right in the eye', to use Francesco Casetti's (1983) expression. This looking at the camera, which is a part of the storyline, breaks the narrative contract with the audience and signals to them the simulated nature of representation each time it occurs. More precisely, it takes the form of a reflexive construction, as Christian Metz described it, which peels off from the storyline to alert the audience that it must read the fiction in the second degree. Narrative complexity demands the complicity of the television viewer and often offers clues.

Tout sur moi... in the middle

Taking what we have said above into account, one might think that *Tout sur moi* would meet with limited public success, given the complexity of its narrative fabric and the risky nature of its contract with the audience. Here we must leave off our analysis for a moment to document the show's ambiguous effect on its audience. The reception by Quebec audiences of this quality television fiction is an exemplary case of a work that the broadcaster wished to drop, but which was saved by its friendly audience, much like some reality shows. Here too the social media played a decisive role. How it unfolded, during the second season of the show in 2008, is of significance to the scholarly field. Three camps were a party to the debate: the broadcaster, the Société Radio-Canada; two critics at the large-circulation daily newspaper *La Presse*; and the public, gathered around a Facebook page. Basically, the arguments went as follows: the public broadcaster, at one end of the spectrum, whose budgets were being increasingly tied to advertising because of government cuts, held up the spectre of insufficient audience ratings to justify putting an end to the show; the two critics, in the middle of the spectrum, described the programme's boldness but indicated that, with a half-million viewers, ratings were indeed low by Quebec standards; and members of the public, right at the other hand of the spectrum, reacted by creating a 'Save *Tout sur moi*!' Facebook page.[2] In the face of this spontaneous public outpouring, the broadcaster backtracked and announced the show's renewal for the third season. Plainly, this anecdote tells us that quality television adapts poorly to the ratings race, even in Quebec, where made-in-Quebec television fiction remains popular; at the same time, however, it shows us that commercial logic can run up against the wall of taste and that interpretive imperatives can be turned into demands for... continuity.

[2] The Facebook page can still be viewed at the address: https://www.facebook.com/groups/9371530846/.

Tout sur moi... in the end

After five seasons of embarrassing situations for the protagonists and repeated soliloquies to the audience, the artists behind *Tout sur moi* undoubtedly thought that the time was ripe to raise the bar even higher. They turned the finale into the high note of the series, in the form of a *trans-textual* tribute to the third-golden-age television series *Lost* (ABC 2004-2010), as proclaimed by the episode's title, *Perdus* (*Lost,* in French). The entire episode is a second-degree television fiction, engaging its audience both in a meta-textual reading of the characters' trajectories throughout the series and in a meta-critical reading of the difference between the show and the American programme. That said, this was, once again, just one case amidst others, and we cannot claim that it represents a change of aesthetic paradigm in step with the earth tremor that was the arrival of quality television. Yet the *trans-textual* quality of the final episode of *Tout sur moi* was not anecdotal; it was generic. It was joined by other third-wave Quebec fictional television programs which also partake of the second degree. These are, in order of appearance: *Grande Ourse* (SRC 2004), which references *Twin Peaks* (ABC 1990-1991); *Les Bougon, c'est aussi ça la vie!* (SRC 2004-2006), which makes winks to *Brutti, sporchi e cattivi* (Ettore Scola 1976), as French scholars have pointed out (Ahl and Frau 2011, 148); *Le coeur a ses raisons* (TVA 2005-2007), which parodies soap operas within fake sets, and with fake wigs, to the point of absurdity; *Musée Éden* (SRC 2010), which dips into a cinematic museum by blending film currents from the silent era of its *diegesis* to offer an audio-visual form that is as coherent with its content as elegant, as Francis Ford Coppola did with *Bram Stocker's Dracula* (1992), notably with German expressionism; and *Apparences* (SRC 2012), which recalls Ingmar Bergman's *Persona* (1966) and its two female characters that are twins more than just by the flesh, a topic on which one of us (Barrette, 2012) has previously written. The finale of *Tout sur moi* is thus the tip of a television iceberg, which, like the deserted island of the television programme to which it refers, can contain secrets or conceal treasures.

Conclusions in the form of a reflexive look back

Theoretical reflection on fictional television programs can take various forms, which depend for the most part on the way one looks at them. To adopt the very clear distinction made by Jean-Pierre Esquenazi (2007), every work follows a trajectory, from its conception to the ways in which it is interpreted, and the theoretical framework employed depends in part on which *moment* of this path the scholar chooses to focus on. The television sociologist Éric Macé (2001, 201) says much the same thing when he states that we must accept the idea that mass culture, and in particular that produced by television, is, like any social object or reality, the 'objectivized' product of an ensemble of mediations inscribed in social relations and whose presence in a specific cultural product (a TV film, a reportage, a commercial, etc.) is never more than a moment in a constant process of the 'configuration' and 're-configuration' of symbolic depictions of the world in which individuals live.

For Howard Becker (1999), for example, one must see 'cultural artefacts' as 'the frozen remains of a collective action', while for Toby Miller (2011, 146), '[one] must realize that programs [do] not fall out of the sky—so we must understand their material conditions of production—and equally, that their meanings are far from explicit and unambiguous, so we must understand their malleable materially as text.' The traditional gap between approaches inspired by semiology and those which derive from a perspective informed by sociology, with on the one side an analysis which says everything about the text without mention of the reality of the mediations it sets in motion and on the other side almost exclusive attention to the work's actors (producers and audiences), is in our view a risky separation in an epistemological perspective. Semiology's aversion to context and sociology's aversion to texts tend to create supposedly independent zones, whereas it is the way in which they mediate, their respective illumination and crossings, that cultural artefacts and their conditions of production and reception take on meaning. As Jean-Pierre Esquenazi (2007, 211) has suggested, it is essential to 'see the work as a process, as a catalyst of encounters, [in order to] be able to compare the various types of social facts to which it gives rise.' For the symbolic practices, we call 'works', far from constituting, as has too often been assumed, islands of meaning, 'unique and sublime' creations operating at the margins of history and society, are on the contrary, in our view, highly social objects whose analysis, evaluation and appreciation must take their dual nature into account. As Esquenazi says in another work (2003, 94), television programmes are 'most of all products whose trajectory one can study as the result of social, economic and political interactions. Next, it is a discourse whose composition should express the phases of its construction.'

That said, from a methodological point of view, it must be acknowledged that textual analyses (analyses of content and representation; semiology) and studies focused more specifically on context (the sociology of 'producers'; economic analysis; the study of reception from an ethnographic perspective) have developed expertise and observational tools, which appear to be largely incompatible and sometimes even contradictory. The solution which consists in seeing the analytical process as one taking place over several stages (such as a textual analysis followed by a reception study), with the goal of confirming or calling into question the existence of a correlation between the two, as seen in the work of David Morley (1982), following that of Stuart Hall (1973), while it has the advantage of not confining the production of meaning either in its textual organization or in its encounter with the empirical viewer, also has the deficiency of being extremely cumbersome to manage, apart from the considerable difficulties posed by the multiple uses of the very concept of reception (Dayan, 1992). We have thus preferred to opt here for a position very close to what is described as *socio-semiotics*, understood, in the words of Andréa Semprini (2007, 15) as 'a discipline which studies the inscription of meaning in social practices' and which we would be tempted to define, in television studies, as the point where *fabula* and *lector* or, if prefer, *poesis* and *polis*, meet. This perspective highlights the complexity of the discourse when this is visible, but also brings out the changes in the way it is perceived when new practices render this necessary.

References

Ahl, N. C. & Frau, B. (Ed.). (2011). *Dictionnaire des séries télévisées.* Paris: Éditions Philippe Rey.

Barrette, P. (2006). De la fiction et des hommes : les séries dramatiques sont l'âme de la télévision de qualité depuis son origine. *24 Images* 128, 38-39.

—. (2008). L'état de la série télé au Québec. Petite enquête en forme de dialogues. *24 Images*, 138, 18-20.

—. (2012). *Apparences,* l'acteur et son double. *Hors-champ*, April 9. Retrieved from http://www.horschamp.qc.ca/spip.php?article489

Barrette, P. & Picard, Y. (2013). Breaking the Waves. In David P. Pierson (Ed.) *Breaking Bad: Critical Essays on the Contexts, Politics, and Style of the Television,* (pp. 121-138), Lanham: Lexington Books.

Baudrillard, J. (1984). *Simulacra and Simulation.* Ann Arbor: University of Michigan Press.

Becker, H. (1999). *Propos sur l'art*. L'Harmattan: Paris. In Maigret, É. & Macé É. (Eds.). *Penser les médiacultures. Nouvelles pratiques et nouvelles approches de la représentation du monde*, Paris: Armand Colin, 2006.

Butler, J. (2010). *Television Style.* London & New York: Routledge.

Buxton, D. (2010). *Les séries télévisées. Forme, idéologie et mode de représentation.* Paris: L'Harmattan.

Caldwell, J. T. (1995). *Televisuality. Style, Crisis, and Authority in American*

Television. New Brunswick: Rutgers University Press.
Casetti, F. (1983). Les Yeux dans les yeux. *Communications*, 39, 78–120.
Dayan, D. (1992). Les mystères de la réception. *Le débat*, 71, 146–162.
Debord, G. (2004). *Society of the Spectacle*. Wellington: Rebel Press.
Dupont, L. (2007). *Téléréalité. Quand la réalité est un mensonge*. Montréal: Presses de l'Université de Montréal.
Eco, U. (1979). *Lector in Fabula*. Paris: Grasset.
Esquenazi, J. P. (2010). *Les Séries télévisées. L'avenir du cinéma?* Paris: Armand Colin.
—. (2007). *Sociologie des œuvres, de la production à l'interprétation*. Paris: Arman Colin.
—. (2003). Éléments de sociologie sémiotique de la télévision. *Quaderni*, 50-51, 89-115.
Gaudreault, A. (2009). *From Plato to Lumière: Narration and Monstration in Literature and Cinema* [*Du littéraire au filmique*]. Toronto: University of Toronto Press.
Genette, G. (2004). *Métalepse: De la figure à la fiction*. Paris: Seuil.
—. (1997). *Palimpsests: Literature in the Second Degree*. Lincoln: University of Nebraska Press.
Hall, S. (1973). *Encoding and Decoding in the Television Discourse*. Birmingham: Centre for Contemporary Cultural Studies.
Jost, F. (2009). *Télé-réalité*, Paris: Le Cavalier Bleu.
Lasch, C. (1979). *The Culture of Narcissism: American Life in the Age of Diminishing Expectations*. New York: Norton.
Létourneau, K. (2010). Présentation. *Argument. Politique, société et histoire*, 13-1, 14-16.
Lipovestky, G. (1983). *L'ère du vide: essais sur l'individualisme contemporain*. Paris: Gallimard.
Macé, É. (2001). Qu'est-ce qu'une sociologie de la télévision? Esquisse d'une théorie des rapports sociaux médiatisés: les trois moments de la configuration médiatique de la réalité: production, usages, représentations. *Réseaux*, 105, 199–242.
McCabe, J. and Kim, A. (Ed.). (2007). *Quality TV: Contemporary American Television and Beyond*. London & New York: I.B. Tauris.
Metz, C. (1990). *FilmLanguage: A Semiotics of the Cinema*. Chicago: University of Chicago Press.
—. (1991). *L'énonciation impersonnelle ou le site du film*. Paris: Méridiens Klincksieck.
Miller, T. (2010). *Television studies: the Basics*. London & New York: Routledge.
Morley, D. (1980). *The "Nationwide" Audience: Structure and Decoding*. London: British Film Institute.
Newman, M. Z. & Levine, E. (2011). *Legitimating Television: Media Convergence and Cultural Status*. London & New York: Routledge.
Perez-Gomez, M. (Ed.). (2011). *Previously On: Interdisciplinary Studies on TV Series in the Third Golden Age of Television*. Sevilla: Frame-Bibliotheca de la Facultad de Communicacion de la Universidad de Sevilla.

Picard, Y. (2009/2010). *Quand le petit écran devient grand*. Montréal: Sur Mesure éditeur.

—. (2011). From Téléroman to Série Télévisée Québécoise: The (Coming of) Golden Age of Quebec TV's Fiction. In Perez-Gomez, M. (Ed.) *Previously On: Interdisciplinary Studies on TV Series in the Third Golden Age of Television*, (pp. 181–195), Sevilla: Frame-Bibliotheca de la Facultad de Communicacion de la Universidad de Sevilla.

—. (2013). De la télé-oralité à la télé-visualité. Évolution de la fiction télévisuelle québécoise du *téléroman* à la *sérietélé* (1953–2012). PhD dissertation, Université de Montréal.

Pichard, A. (2011). *Le nouvel âge d'or des séries américaines*. Paris: Le Manuscrit.

Récanati, F. (1979). *La transparence et l'énonciation. Pour introduire à la pragmatique*. Paris: Seuil.

Semprini, A. (ed.) (2007). *Analyser la communication 2. Comment analyser la communication dans son contexte socioculturel*. Paris: l'Harmattan.

CHAPTER FOUR

MUHTEŞEM YÜZYIL OR MUHTEŞEM REZALET: CONTROVERSY SURROUNDING THE TELEVISION SERIES *MUHTEŞEM YÜZYIL* AND THE CRISIS OF TURKISH IDENTITY

CHERIE TARAGHI

Media hype, criticism, and controversy surrounded the Turkish TV series *Muhteşem Yüzyil* (Magnificent Century) even before the first episode was aired on 5 January 2011, and continue to this day. Turkish media of all sorts frequently report and discuss aspects of the series, from developments in the narrative storyline and comments on the series' stars to much stronger criticism regarding the very existence and premise of the series: its subject, the life of Ottoman Sultan Suleiman and his relationship with his eventual wife Hurram within the Harem of Topkapı palace. The series has struck at the core of debates on Turkey's history and identity, exposing all the taboos and ideological strangleholds that make up the divisive faultlines of Turkey's heterogeneous population.

At the risk of simplifying Turkey's complex culture and the multi-faceted identity struggles within the country, this article focuses more specifically on the criticisms and debates inspired by the series that have been most centre stage and pertinent in Turkish political and media circles: the struggle between the traditional urban Kemalist elites of Turkish society, with their secular, modern and Westernized identity, which rejects the Islamic, Ottoman past and identifies with a Central Asian Turkic identity, together with the liberals and secular segments of Turkish society, and the Conservative Nationalist segments of Turkish society, who form the current Development and Justice Party (AKP) government and Prime Minister Recep Tayyip Erdoğan's electoral stronghold. The Conservative Nationalists are made up mainly of conservative, pious Turks from the peripheral regions of the country, who form an identity based on the strength and triumph of Turkey's Sunni Muslim Turkish Ottoman past and a neo-Ottoman desire to be powerful and respected on the world scene once more.

The article concentrates on two phases of the controversy surrounding the series *Muhteşem Yüzyil*; the initial uproar before and just after the series was first aired in January 2011 and the reaction after Prime Minister Erdoğan strongly criticized the series in November 2012.

Turkish Television Series

The history of television series in Turkey commenced with a series called *Kaynanalar* (The Mother-in-Laws), which was broadcast for 32 years. The 1975 adaption of the turn-of-the-century Turkish novel *Aşk-ı Memnu* (Forbidden Love) was the first series which glued audiences to their TV screens. Over the years, a large number of very popular series with extremely high rating/ viewership were produced, namely *Kartallar Yüksek Uçar* (Eagles Fly High) and *Kuçuk Ağa* (Little Man). However, since the early 2000s, Turkish TV drama has experienced such unprecedented growth and popularity that it has surpassed all other forms of prime-time TV programs, particularly when one considers the number of series available any night of the week across a broad range of channels.

The lifespan of a TV series is determined by its popularity and ratings. The various TV channels compete with one another for ratings. It not uncommon for different channels to pit series with interesting or controversial storylines or starring various famous and popular stars against one another in order to influence audience loyalty. For example, in September 2011, the channel Kanal D tried to counter the popularity of the series *MuhteşemYüzyil*, which is shown on the channel Star and was first broadcast in January 2011, by placing its new promising series *Kuzey Guney*, starring the extremely popular Kıvanç Tatlıtuğ, in exactly the same time slot as *MuhteşemYüzyil*.

The majority of mainstream Turkish TV series correspond to the international genre known as 'soap opera'. The series tend to be melodramatic, dominated by a romantic storyline, forbidden and interesting love stories, with a focus on the inter-relationship, intrigue, love, and conflict between family members. On the whole the series reflect the lives of urban, middle and upper class characters who live in gorgeous mansions and lead luxurious lives and most series represent various social issues current in the country.

MuhteşemYüzyil: Basic Summary of the Series

The series *Muhteşem Yüzyil* is a perfect example of a mainstream soap opera; it has an episodic structure and an ongoing narrative. Each episode is a more or less complete set of events containing crises and resolutions. However, each episode is embedded within a larger narrative framework that essentially visualizes life within the Harem of the tenth Ottoman Sultan, Suleiman the Magnificent, during the period of his reign.

The narrative is chronological, commencing at the point of Sultan Suleiman's ascension to the throne at the age of twenty-six upon the death of his father, Sultan Selim I. Limited attention is paid to the political aspects of Suleiman's reign. A number of episodes and scenes are dedicated to some of his battles and expeditions and there are occasional scenes where foreign diplomats are received or court politics are discussed and various Vezirs and Paşas debate issues or scheme, adding to the spice and intrigue of the programme. Most of the two-hour screen time reflects life within Suleiman's harem, his life-long love affair with Hurram, her political machinations to retain the Sultan's favour and ensure one of her sons becomes heir to the throne and in general, the rivalry between the Sultan's many concubines and the conflict between his family members.

Stage One: the conservative criticism

Controversy surrounded the series even before its first episode was broadcast on TV on 5 January 2011. Initial trailers advertising the imminent arrival of the new series were criticized and many members of the public complained to the Turkish Radio and Television Supreme Council (RTUK) that the series misrepresented and abused the image of the historical figure, Sultan Suleiman, due to repetitive depiction of the Sultan with women in the harem and of the Sultan holding what appears to be a wine goblet in his hand. Complaints continued to pour in after the first episode was broadcast.

RTUK's reaction to these complaints was to warn the series' producers that the show was an 'affront to the society's national and moral values' since 'it did not show necessary sensitivity to the privacy of a historical figure.' As a result the producers and the scenarist made changes to the second episode (and all other episodes thereafter), especially regarding scenes with kissing / alluded sexuality, in order to express the required sensitivity to the Sultan's 'privacy'. Complaints have continued to pour in throughout the three seasons that the series has been shown on television. As a result of 'repeated' offenses, RTUK has to date fined the TV channel

782,771 TL.[1]

Complaints, fines, and threats of removing a series/ program from television are quite common in Turkey and by no means unique to *Muhteşem Yüzyil*. In deference to to the concept of national and moral values (Milli ve Manevi Değerler), various series such as *Kurtlar Vadisi* (Valley of the Wolves) have even been removed from TV (2007), due to excessive violence and the politics they expressed. Others such as the remake of *Aşk-I Memnu* were by official request studied and although the series was not removed from TV, in the words of the media 'a fatwa was passed'[2] against it (2012) due to its incongruous stance with regards to religion (dine aykırı), for showing sexual relations between an unwed couple and abortion, among other offenses. What is special about the debate regarding *Muhteşem Yüzyil* is the fact that this is the first time a historical costume television series is being broadcast and it is the first time the life of a time- honoured and highly revered historical figure such as Sultan Suleiman is being portrayed on a weekly basis.

On one level the issue that is under debate is the fact that the producers and initial scenarist Meral Okay opted to portray the Sultan's 'private' life; to focus on life within the harem rather than the Sultan's political deeds and battles to expand the power and geographical area under the control of the Ottoman Empire. In an interview on national television after the airing of the first episode, Okay stated that the decision to focus on the Sultan's private life was deliberate in order to make the series more accessible to and entertaining for viewers. In her view, audiences would not be interested in a documentary-style series that would focus on precise historical details about Sultan Suleiman, the leader, the hero, the warrior, the lawgiver, with no information regarding Suleiman the man. Okay makes it very clear that the series is in fact 'fiction' (kurgu) inspired by historical figures and intended to be entertainment rather than a history lesson.[3] In her own words, 'We have been saying the same thing from the start: this is a fiction inspired by history. By entering the harem, we made all those untouchable and respected characters of history closer to us. We gave them a material existence as humans, with fears, anger and passions.' (Fowler, 2011).

The scenarist has two historical advisors and prepared for two and a half years prior to writing the scenario by reading every document she

[1] Muhtesem Yuzyil'a Muhtesem Ceza!, *Haber365*, 24 April 2013.
[2] See for example, Ali Ekber Erturk, 'Aski Memnu'ya RTUK Fetvasi', *Aksam*, 7 December 2012.
[3] Interview with Meral Okay by Banu Guven on *ARTI* program, *NTV*, 6 January 2011.

could find on sixteenth-century Ottoman history. Every effort is made to pay attention to detail and ensure that the images shown, the buildings, decorations, market area and streets, and so forth are accurate representations of the historical time frame the series represents. In short, Okay states that she and the producers of the show have no intention of insulting history or debasing the image and honour of the Sultan or the Ottoman Empire.[4]

Regardless of intentions, the fact is that history in Turkey is ideological and the sixteenth century, the Golden Age of Ottoman power and prestige, together with Sultan Suleiman who ruled during that century are particularly revered by the Conservative Nationalist (Muhafazakar Milliyetçi) segments of Turkish society; this segment more or less corresponds to the cross section of society that votes for the ruling Justice and Development Party (AKP).

Fears of the Conservative Nationalist Segment

One of the principal fears and complaints put forward by the Conservative Nationalist segment of Turkish society is that Sultan Suleiman's image, dignity, and reputation are ruined by the content of the series and that this generation of Turks is learning 'history' through the fallacies shown in the series. A simple but very revelatory example is a comment by the columnist Sema Karabiyik in her column for the conservative, religious (Islamic) newspaper *Yeni Şafak.*[5] Writing in wake of the initial criticisms made against the series in January 2011, she states that as a result of the series, 'All of a sudden in the newspaper columns, on internet sites, for a large number of families and children, lawful Sultan Suleiman has turned into 'Suluman' (a reference to the scene in the second episode of the series where the character Hurram, in her supposed Ukrainian accent, pronounces Suleiman's name as 'Suluman', before passing out in his arms), and has become excessively frivolous. If this is what is intended then they (the producers) have hit the mark very early on; I congratulate them. I would like to point out that the language used to criticize the series in the news is not harming the series; rather the harm is to Sultan Suleiman' (Karabiyik, 2011). The example offered in this article is a minor but apt example of the fears experienced by the Conservative

[4] Interview with Meral Okay by Banu Guven on *ARTI* program, NTV, 6 January 2011.

[5] It must be pointed out that *Yeni Şafak* newspaper is one of the prime voices for the Conservative Nationalist segment of society. Another important newspaper bearing their voice is *Vatan Gazatesi*.

Nationalist segment of Turkish society.

The official image of Suleiman, which is also taught in national curricula of history textbooks, is that of a 'pious, compassionate, kind, generous, humane, and modest man and a discreet, far-sighted, balanced, law-abiding, and just ruler. He is considered to be an independent decision-maker. All these qualities correspond to those of a perfect ruler enumerated in the medieval mirror-for-princes literature. His most important quality is that he was just and to this effect law-abiding. He is praised as being uncompromising against all illegal, oppressive acts committed by the powerful against the powerless. Suleiman is idealized as a promulgator of just laws and as a strict administrator of his laws. He succeeded in promoting such an image for himself. Suleiman is given the impression, or the myth is created of, a perfect ruler' (Inalcik , 1992). He is also portrayed as a warrior, a leader who spent most of his life in battle to expand the land mass ruled by the Ottoman Empire.

When one takes this into consideration, it is easy to understand why segments of Turkish society, especially with their religious sensitivity and conservative social and cultural values, are incensed and threatened by the thought that theimage of Suleiman presented in the series, with all the dazzling costumes, jewellery and set decorations, will replace the 'history book' Suleiman. Sultan Suleiman is suddenly transformed from the 'perfect ruler' into a 'man': a man with passions, a strong libido, who spends a lot of his spare time chiselling gems to make jewellery for his favourite female of the moment, who makes mistakes and whose decision-making, far from being totally independent, can be influenced by the people around him, particularly the women in his harem.

History and Identity as Ideology

Reactions against the series can be better understood when one takes Republican Turkish history into consideration. As the historian Esra Özyürek puts it: 'In 1923, the newly founded Turkish Republic committed to a modernist future by erasing the memory of its immediate Ottoman past. […] The new regime established itself as a homogenous and secular nation-state that rejected the multicultural heritage of the Ottoman Empire and its emphasis on Islam' (Özyürek, 2007). The Republic that was founded after the victory in the 'National Struggle' organized after the collapse of the Ottoman Empire does not merely imply a regime change but also the imposition of a new ideology and construction of a totally new modern, Westernized, and secular identity, which has come to be known as 'Kemalism'. Through decades of public education the Turks have been taught that Kemalism or being a Turk means: 'We are a Turkish race, of Turkish blood, of Muslim faith; our historical roots lie in Turkic Central Asia; we believe in Ataturk's project of modernization; we are secular (which in Turkey means a secular lifestyle within a system of state sponsored Sunni Islam)' (White, 2012, p. 6).

The secularization campaign of the early republican years lost steam by the 1950s, and interest in Islam and the Ottoman Empire made a gradual comeback to the public sphere and politics. The 1980 military intervention was a particular impetus as the military encouraged more freedom for a modest, state-defined form of Islam to combat the appeal of communism and leftist ideologies. Religious education was made compulsory for all in schools. Quran classes were opened, and state-controlled moral and religious education was promoted. This 'Turkish-Islamic synthesis', an ideology built out of Ottoman, Islamic, and Turkish popular culture, was used to unify and stabilize the power of the state.[6]

The first elected party, the Anavatan Partisi's (Motherland Party) government after the 1980 coup, opened Turkey's insular, state-led economy to competition in the world market. This resulted in the growth of many small and medium-sized businesses in the provinces that were owned by pious Muslims. Their wealth created a market for Islam-friendly bourgeois products and lifestyles (an Islamic economic sector) and initiated a Muslim cultural renaissance in fashion, lifestyle, leisure activities, novels, media, and music. The pious elite's wealth also supported overtly Islamic politicians and their programs throughout the 1980s and 1990s. One beneficiary is the AKP, a party which represents this Islamic power base.

The growth of this conservative, pious elite has seriously undermined the social and political leadership of the secular urban Kemalists. Kemalist control of the educational system and urban economic and social life had provided and promoted an orthodox national identity. The emblematic citizen was a Turkish Muslim with a secular lifestyle, dedicated to a state-led program of modernization believed to be Ataturk's design. Since coming to power, the AKP and its pious supporters, in contrast, have

[6] It is important to note that while religion, that is secular versus religious segments, is the prime source of political cleavage and conflict in contemporary Turkey, conditions in the country are in fact more complex. However, Kemalism versus the religious Conservative Nationalists do form the two most vocal and visible cleavage blocs in the country and that is why this article focuses only on these two segments of Turkish society, particularly since the controversy surrounding the series *Muhteşem Yüzyil* primarily exemplifies the arguments and fears of these two segments.

developed and implemented an alternative definition of Turkishness and the nation; one that imagines Turkey not as a nation embattled within its present political borders but as a flexibly bounded Turkey that is the self-confident successor to the Ottomans in a rediscovered (and reinvented) past. This Turkish identity, the Conservative Nationalist one, is that of a pious, conservative Muslim Turk whose subjectivity and vision for the future is shaped by an imperial Ottoman past overlaid onto a republican state framework, but divorced from the Kemalist state project (White, 2012, p. 9). In fact, the AKP has gradually brought all sources of traditional Kemalist bureaucratic opposition under its firm control, including the judiciary, the military, and the civil service. Instead of commemorating the 1923 founding of the Turkish nation, the Conservative Nationalists pay tribute to historical events like the 1453 Ottoman Muslim conquest of Christian Byzantium. If symbolic representation of this new identity were necessary, then Istanbul's newest cultural attraction, the municipal-run Panorama 1453 History Museum, located just outside the city's ancient walls, which tells the story of the Ottomans' conquest of Byzantine Constantinople would fit the bill.

The Kemalists and the Conservative Nationalists view each other with extreme suspicion. The latter suspect that the former are always ready to subvert the democratic process, for example to unleash another military coup or to use the judiciary to have the AKP banned and removed from power; and the former believe that the latter are subverting democracy through imposing increasingly authoritarian, religious laws and are out to turn Turkey into an Islamic state.

Since the two sides of the religious cleavage in the country view their positions as a zero-sum situation, the strong reaction against the series *Muhteşem Yüzyil* is easy to comprehend. For the Conservative Nationalist segments of society the series is a direct affront, a deliberate effort to insult and denigrate the image and identity of the shining hero of their historical identity, Sultan Suleiman. Within the highly ideological context of Turkish society, the insult is not considered to be levelled against the identity of a historical figure alone; it is a conspiracy to insult them: their identity, their values and beliefs. The series exploits many of the taboos of conservative, religious Turkish society, such as extra-marital sexuality, fixation with sex, the habit of 'hanging out' at the brothel, consumption of alcohol, revealing, un-Islamic manners of dress and demeanour, and so forth.

Even before the first episode was shown on TV, the series was labelled as 'Muhteşem Rezalet' (Magnificent Disgrace). A few examples, from different types of media, of the reaction by the Conservative Nationalists will be detailed below to provide more concrete demonstration of the sentiments and worries of this segment of the society vis-à-vis the series.

A number of Conservative Nationalist writers, columnists, and historians, such as Mehmet Niyazi and Yavuz Bahadiroglu spoke out against the series and condemned the content. In a news item about the upcoming series, broadcast on the Islamic Samanyolu TV channel on 4 January 2011, Mehmet Niyazi states, 'It's lies. It is an effort to discredit the Ottoman Empire or more correctly to blacken our history.'[7]

An article entitled 'Not Magnificent Century but Magnificent Disgrace and a Betrayal of History', published in the website Diyanethaberler.com on 10 January 2011, accuses the producers of 'purposefully warping historical truths', 'placing a wrong message in the people's subconscious', 'placing obstacles and barricades in front of young people so they may not know the truth about their ancestors', 'belittling a leader who ruled the world for half a century', 'aiming to make the young people of the current generation embarrassed about their own history and past', 'opening the path for young people to have a complex and hate their ancestors and experience an identity crisis', and so forth.[8]

Another example of the Conservative Nationalist reaction to the series is the article 'Magnificent Betrayal' by *Yeni Akit* newspaper columnist Yener Dönmez. Dönmez commences his column of 8 January 2011 in the following manner: 'Just as the 'Sick Man' was beginning to get well, a period where its old grandeur is beginning to be spoken about… at a time period where Turkey's name is being mentioned with respect from the Middle East through to the Caucuses, from the Balkans through to Africa, from America through to the South East, that is throughout the world, such a TV series appears…' The crux of the column is that the series is a 'betrayal' and an insult by liberal Socialists to the new 'neo-Ottoman' status of the country (Dönmez, 2011).

On the other side of the spectrum, on the whole the Kemalists reacted to the Conservative Nationalists' criticisms rather than to the series itself. Few actually defended the series or its contents per se. Historians such as Ilber Ortayli mentioned a few historical inaccuracies but overall praised the series. The influence of the series *The Tudors* as a source of inspiration in terms of look and content (focus on private life of the Sultan) was also pointed out.

Columnists for the left leaning, Kemalist newspaper *Cumhuriyet*, such asBekir Coşkun, Zeynep Oral, and Deniz Kavukçuoğlü or Can Dundar

---

[7]'Muhtesem Yuzyil gercegin tam tersi', *Samanyolu TV*, 4 January 2011.

[8] 'Muhtesem Yuzyil degil, muhtesem rezalet ve tarihe ihanet!'. *diyanethaberler.com*, 10 January 2011.

from the more liberal *Milliyet* comment on the 'neo-Ottoman' stance of the government and the Conservative Nationalists, criticizing this as an affront to Ataturk's vision for the country. There is also a tendency to point out that these 'neo-Ottomans' have few qualms when Ataturk's name or memory is insulted, but they become 'sensitive' regarding 'national values and moral' when their Ottoman hero's memory is under review. There is also a tendency to use the reaction of the Conservative Nationalists in order to criticize the AKP government and its 'prohibitive' policies and 'authoritarian' stance, demanding more tolerance and acceptance of other views. Or to heighten Conservative Nationalist worries by emphasizing all the negative aspects of the Ottoman Empire, such as murder of children and siblings in order to avoid coups, naming all the Sultan's that drank alcohol or mentioning the fact that none of the Sultans ever went to Mecca for Haj and so forth.

Some take a somewhat softer tone, pointing out, as the producers do, that the series is based on history and historical figures but the storyline is fiction. And another liberal stance was to point out that it is mistaken to treat leaders and historical figures as gods; be it Sultan Suleiman or Ataturk, these individuals are 'human', with the human tendency for good and bad, successes and failures, weaknesses as well as strengths.

A comparison has repeatedly been drawn between *MuhteşemYüzyil* and the journalist, columnist, and documentarian Can Dundar's documentary *Mustafa*, produced in 2008. *Mustafa* is a documentary about the life of Mustafa Kemal Ataturk. The film depicts Turkey's national hero as a flawed man who drank heavily and suffered from bouts of loneliness. It presented him as a dictator and tried its best to portray a realistic representation of a 'man' with weaknesses and strengths, as opposed to the 'divine' figure Turks are indoctrinated to worship and never criticize. The parallels between *Mustafa* and *MuhteşemYüzyil* are quite clear. However, while the latter shakes the foundations of Conservative Nationalist veneration, the former is considered an insult and betrayal by the Kemalists. Worst of all, it is a betrayal for the Kemalists by one of their own, as Can Dundar has clear Republican credentials. In both cases the products have been controversial and have been criticized for the same reason: an inability and need to come face to face with the realities of history; learning to recognize that the leaders depicted are human and that portraying and discussing the truth about their characters or the historical conditions they lived under does not detract from them and what they have achieved. It is valuable to note that the reaction of the Conservative Nationalists to the documentary *Mustafa* and the controversy raised by the Kemalists was the exact reversal of the situation faced with

MuhteşemYüzyil. Neither side was able to learn from the experience of the other or sympathize with the ideological fears and worries the other faced when their values and sense of morality were brought under question.[9]

Stage Two: Erdoğan's Criticism and Liberal Defence

The controversy surrounding the series *MuhteşemYüzyil* gained momentum once more on 25 November 2012 when Turkish Prime Minister Recep Tayyip Erdoğan criticized it during a public speech. The speech in fact does more than criticize the series, it is also a priceless example of the sense of worry the series invokes in the Conservative Nationalist segment of society. It is also indicative of Erdoğan and the AKP's 'neo-Ottoman' ideology and increasingly authoritarian stance. Erdoğan states, 'We will go everywhere that our ancestors have been to… People, I think, know our ancestors as those portrayed on the documentary *MuhteşemYüzyil*. We don't have such an ancestor. We are not aware of such an ancestor. We are not aware of such a Sultan Suleiman. Thirty years of his life were spent on horseback. They were not spent in the palace as they portray in the series… I condemn the producers of the series as well as the owners of the television channel here in front of our citizenry. Although I have warned all those involved, I also expect the judiciary to take the necessary steps to decide on the series. Such a perception is not acceptable. Those who play with this nation's values need to be given a lesson as a nation; as a nation they need to be answerable to the dictates of the law.'

Erdoğan's criticism was quickly followed up by AKP Istanbul Deputy Oktay Saral drafting a parliamentary bill in an attempt to ban dramas that 'distort or change historical realities' and depict them differently. Saral's bill suggested making an addition to a clause in Article 8 of the Radio and Television Supreme Council (RTÜK) law, which states that media service providers are prohibited from encouraging crimes, praising criminals and criminal organizations, and providing information on how to commit a crime. The proposed addition is as follows: broadcasted programs are banned from 'insulting, humiliating or distorting' historical events that are accepted in accordance with society's national values and the widely

---

[9] For an interesting example of the debate on both the series and the documentary and the comparisons drawn between the two in the media see the episode of the talk show *32. Gun* aired on 3 December 2012, which included Can Dundar himself. The episode is available at
http://www.youtube.com/watch?v=GVvP13EtCkY

accepted facts about historical personalities.[10]

The Conservative Nationalist worries and concerns regarding the way Sultan Suleiman is portrayed in the series, as well as their fear that the series is working to change the understanding and respect the citizens in general have of the Ottoman Empire, Turkey's glorious ancestry, has already been discussed in some detail. This speech brought forth a new dimension in the controversy. While many citizens had complained to RTUK and demanded the removal or cancellation of the series from television, no actual step had been taken in that direction. Erdoğan's demand that judicial action be taken opened the path for such a possibility, as well as extensive speculation regarding the future of the series.

In terms of the series itself, Erdoğan's comments resulted in the series becoming visually more pious and conservative. From the speech onwards efforts were made to ensure the female characters cover their hair when in public or when facing a male. The costumes in general also became more conservative, showing less cleavage.

In many ways Erdoğan's criticisms turned the tables, and this time it was the Kemalists and liberal segments of Turkish society that were incensed. As with the initial controversy, the reaction Erdoğan's comments generated resulted in defence of 'art', 'free speech', 'values and morals', the demand for tolerance of all lifestyles and points of view, the issue of 'identity' and 'ancestry', as well as the need to face history. The series was the symbol whose meaning everyone discussed, rather than discussing the series itself.

Erdoğan's reference to the ancestor, embracing the Ottoman past and a Sunni Muslim Turkish identity, met with a very strong reaction among all secular, liberal, and Kemalist segments of Turkish society. As already pointed out the notion of ancestry has been contentious since the inception of the Turkish Republic when the Kemalists harked back to the Central Asian Turkic roots and rejected Ottoman and Byzantine pasts, to inspire pure Turkish identity. The criticism encouraged some discussion and argument regarding ancestry and the need to stop inventing a homogenous, reductionist Turkish history and identity, which excludes segments of Turkey's highly heterogeneous population.

Erdoğan was accused of distorting history by reducing Ottoman identity to that of the Sunni Muslim Turk. One popular reaction was to attack Erdoğan on an issue where the Conservative Nationalists had already expressed considerable weakness. It was pointed out that, as the

---

[10] 'Deputy proposes ban on TV dramas 'distorting' history', *Today's Zaman*, 5 December 2012.

series portrays, the Ottoman Sultans did have harems and most of the women in the harem were Christian concubines, so in fact the mothers of most of the Ottoman Sultans were Christian. A specific criticism which makes the ideological division regarding 'ancestry' and 'identity' in Turkish society very clear is the criticism made in Parliament by the Kurdish Bariş ve Demokrasi Partisi (Freedom and Democracy Party) Dıyarbakır Deputy, who on 17 December 2012, rhetorically asked Erdoğan, 'My brother, why did *your* ancestor set up this harem? Was this harem a child protection centre? Was it a dormitory Quran Course for girls? Why where there no Muslims in this harem? Why were *your* ancestors against official marriage?'[11]

The Conservative Nationalists have also been asked to recall that the Ottomans conquered Byzantine land and in embracing a Muslim Turkish identity, the Byzantine past was being erased, together with the fact that the lands of Anatolia, presumably home of the Turks, were populated with people from all sorts of ethnic backgrounds such as the Armenians, Kurds, Circassians and Greeks. This was one of the more interesting aspects of the reaction to Erdoğan's criticism as it in fact forced the staunch Kemalists to also face and question the 'official' Republican history.

The identity of the pious Halife-Sultan was also taken under the microscope, pointing out various Sultans who drank alcohol and even died of alcohol-related causes, the fact that no Sultan ever went to Mecca, that Fatih Sultan Mehmet had an Italian painter paint 'pornographic' painting on the walls of Topkapı palace, and so on.[12]

The need to face the reality of history as opposed to an idealized myth, which cannot be criticized or questioned, re-surfaced in December 2012. This time the focus was no longer on the inviolability of the identity and character of Sultan Suleiman. Almost two years after the commencement of the series, the narrative was fast approaching uglier aspects of Sultan Suleiman's rule; for example, the fact that he orders the execution of his grand vizier Ibrahim Pargali, and the execution of his eldest son, Mustafa, and of one of his younger sons, Beyazid. What was now questioned was how PM Erdoğan and the conservative nationalists would face and accept these ugly historical truths, the 'blackening' of the image of the idealized

[11] 'Altan Tan'dan Muhteşem Yüzyıla Destek', *Internet Haber*, 17 December 2012.

[12] For examples see Ahmet Altan, 'Ecdat', *Taraf*, 28 November 2012; Ahmet Hakan, 'Ve Nihayet Çaktı Muhteşem Yuzyıla'. *Hurriyet*, 26 November 2012; Mine Şenocaklı, Interview with Murat Belge, *Vatan*, 10 December 2012.

hero. Would the scenarists be allowed to portray these facts and how?[13] Clearly if Saral's proposal was made law then this question would become very pertinent indeed.

With the arrival of the draft law the issue was elevated beyond the status of one TV series and took on national importance. The interpretation of history is highly subjective after all. The hero of one segment of society may well be despised by another. As one Turkish TV scenarist asked with reference to more recent controversial Turkish history, 'Who will decide which historical figure represents national values? When describing Sultan Abdulhamit II do we have to discuss the oppressive environment and his universally feared secret police or not? What about Enver Paşa and the Armenian deportation issue? And when discussing the Republic's second President İsmet İnönü do we have to talk about the massacre of Dersim? Does discussing these truths hurt these historic personalities or not?'[14]

All of this has certainly focused attention in Turkey on the issue of media freedom and freedom of speech and the place of artistic freedom of speech within the context of Turkish democracy. Criticism of Erdoğan's stance, his call for judicial action, and the government's rush to fulfil this call has poured in from all segments of secular and liberal Turkey. The most common critique projects the Kemalist worry that Erdoğan and his government are interfering in too many facets of the lifestyle they cherish.

PM Erdoğan has over the years commented and criticized, in the name of Turkish values and morality, on all aspects of life from television to art to the way people live their lives, the issue of abortion and caesarean operations as an option for child birth, the number of children a Turkish family should have, consumption of alcohol, public demonstration of affection, and so forth. Increasingly Turks of liberal inclination have become frustrated with this constant stream of interference and attack on their way of life. In the zero-sum context, all this has added to the sense of a growing Islamic authoritarianism bound to trample the secular 'way of life'. The June 2013 'Gezi' protests in Turkey were exactly about this issue. The decision by the government, firmly backed by the PM Erdoğan, to destroy Gezi Park and rebuild a nineteenth-century army barracks in which a hotel, shopping centre, etc could be incorporated was the spark from which erupted the fire of anger and disdain for the PM, the government, and its policies. Criticism of the series *Muhteşem Yüzyıl* was another spark. A comment in an interview with television series director

[13]For examples of articles and comments along this line see Melih Aşik, 'Suçlar ve Cezalar', *Milliyet*, 12 December 2012; 'Tarihi Kriteri Kim Belirliyecek', *Radikal*, 6 December 2012.

[14]'Tarihi Kriteri Kim Belirliyecek'. *Radikal*, 6 December 2012.

Kudret Sabancı in January 2013 sums up this sentiment very well. Talking in fact about Erdoğan's criticism of *Muhteşem Yüzyıl*, Sabancı states, 'Politics has entered as far as our very bedrooms' (Bahar, 2013).

In an article on Erdoğan's speech in his column in the newspaper *Cumhuriyet*, sociologist Emre Kongar typifies this sentiment of the liberals and the seculars which burst at the Gezi Protests. 'The PM's stance and AKP's suggestion (Saral's proposed law) totally defines one man rule; it corresponds to the understanding of a dictatorship. With this stance, neither democracy, nor freedom, nor rule of law remains. Neither does the ascendency of law remain, nor a separation of powers… The issue is not one of a TV program; it is not the issue of the personality of the Prime Minister; it is the issue of a regime… It is the issue of a regime that is transforming from a democracy into a dictatorship' (Kongar, 2012).

Conclusion

During the past three years the series *Muhteşem Yüzyil* has sparked a great deal of debate and controversy in Turkey; while on the surface these debates are about the series, its subject and contents, in reality the controversy has embraced identity struggle at the very core of Turkey's contentious modern history. The series has resulted in a political and media volley of exchanges about whose vision and ideology in fact defines and encapsulates the identity of present day Turkey: is it that of the urban, traditional Kemalist elites, with their modern, secular, Westernized views and lifestyle or that of the Conservative Nationalists, led by the current Turkish PM Erdoğan, espousing the pious, Sunni Muslim Turkish identity which identifies with the glories of the Ottoman Empire and Ottoman Sultans as the source of its ancestry. As the paper shows, the debates and controversy in fact expose the non-embracing fault lines of Turkey's current identity crisis and the zero-sum situation the struggle between these ideological views of history and identity has produced.

References

Ahmet, A. (2008). Can Dündar'ın 'Mustafa' filmi tartışmasına Ahmet Altan da katıldı. *Zaman Gazetesi*, 24 November. Retrieved from http://www.zaman.com.tr/gundem_can-dundarin-mustafa-filmi-tartismasina-ahmet-altan-da-katildi_756716.html

Akdeniz, E. (2010). Despite Criticism at Home, Turkish TV Series Continue Shaking World. *Today's Zaman*, 2 May. Retrieved from http://www.todayszaman.com/news-209049-despite-criticism-at-home-turkish-tv-series-continue-shaking-world.html

Akyol, M. (2008). Turkey's Liberal Islam and How It Came To Be. *Turkish Policy Quarterly*, January 2008. Retrieved from http://www.turkishpolicy.com/images/stories/2008-01-turkey/MustafaAkyol.pdf

Ateş, A. (2013). Occupy Gezi, beyond the religious-secular cleavage. *SSRC*, 10 June 2013. Retrieved from http://blogs.ssrc.org/tif/2013/06/10/occupy-gezi-beyond-the-religious-secular-cleavage/

Bahar, D. (2013). Interview with Kudret Sabancı. 'Siyaset Yatak Odalarımıza Kadar Girdi', *Milliyet*, 2 February 2013. Retrieved from http://cadde.milliyet.com.tr/2013/01/02/HaberDetay/1649866/-siyaset-yatak-odalarimiza-kadar-girdi-

Çağaptay, S. (2006). *Islam, Secularism, and Nationalism in Modern Turkey -Who is a Turk?* Routledge, London.

Çağlar, K. (2013). First Person Singular. *London Review of Books Blog*, 3 June 2013. Retrieved from http://www.lrb.co.uk/blog/2013/06/03/caglar-keyder /first-person-singular/

Çiğdem, B. (2011). The Soft Power of Turkish Television. *South East European Times*, 23 July 2011. Retrieved from http://www.setimes.com/cocoon/ setimes/xhtml /en_GB/features/setimes/features/2011/07/23/feature-02

Cizre, U. (2008). *Secular and Islamic Politics in Turkey: The Making of the Justice and Development Party*, Routledge, London.

Deputy proposes ban on TV dramas 'distorting' history. *Today's Zaman*, 5 December 2012. Retrieved from http://www.todayszaman.com/news-300267-deputy-proposes-ban-on-tv-dramas-distorting-history.html

Dizileri Yayindan Kaldirmak Icin. *Vatan Gazetesi*, 16 April 2013. Retrieved from http://haber.gazetevatan.com/dizileri-yayindan-kaldirmak-icin/538742/11/medya

Ekber, E. A. (2012). Aski Memnu'ya RTUK Fetvasi. *Aksam*, 7 December 2012. Retrieved from http://www.aksam.com.tr/guncel/ask-i-memnuya-rtuk-fetvasi--152833h/haber-152833

Esra, Ö. (2007). *The Politics of Public Memory in Turkey*, New York: Syracuse UP.

Fowler, S. (2011). The Dirt and the Soap on the Ottoman Empire. *New York Times*, 17 March 2011. Retrieved from http://www.nytimes.com/2011/03/17/world/middleeast/17iht-m17-soap.html?pagewanted=all&_r=1&

—. (2011). Magnificent Century Divides Turkish TV Viewers over Life of Suleiman. *The Observer*, 20 March 2011. Retrieved from http://www.guardian.co.uk/world/2011/mar/20/turkish-drama-magnificent-century-divides

Halil, I. (1992). Sultan Suleyman: The Man and the Statesman. In Gilles Veinstein (Ed), *Soliman le magnifique et son temps*, La Documentation Francaise, Paris.

Ilker, A. (2012). Uyardik ama Yargi. *Vatan Gazetesi*, 25 November 2012. Retrieved from http://haber.gazetevatan.com/uyardik-ama-yargi/495200 /9/siyaset

Jones, D. (2012). 'Turkey's PM attacks popular Show', *VOA News*, 12 December 2012. Retrieved from http://www.voanews.com/content/turkey-tv-show-erdogan/1563806.html

Karabiyik, S. (2001). Dizilerden Tarih Ogrenilmez. *Yeni Safak*, 23 January 2011. Retrieved from http://yenisafak.com.tr/yazarlar/SemaKarabiyikPazar/dizilerden-tarih-ogrenilmez/25797

Kongar, E. (2012). Muhteşem Yüzyıl'in Sorunu Değil, Rejim Sorunu'. *Cumhuriyet*, 29 November 2012. Retrieved from www.cumhuriyet.com.tr/?hn=381726

Mehmet, B. (1997). Yasin Yağcı, 'Tarih Ekranda Yargılanıyor'. *Aksiyon*, 22 November 1997. Retrieved from www.aksiyon.com.tr/aksiyon/haber-3203-34-tarih-ekranda-yargilaniyor.html

Melih, A. (2012). Suçlar ve Cezalar. *Milliyet*, 12 December 2012. Retrieved from http://gundem.milliyet.com.tr/suclar-ve-cezalar/gundem/gundemyazardetay/12.12.2012/1640620/default.htm

Mert, N. (2011). Muhafazakarlarin Kanuniya Sevmeyen Kahramani. *Milliyet*, 29 January 2011. Retrieved from http://www.milliyet.com.tr/muhafazakarlarin-kanuni-yi-sevmeyen-kahramani/nuray-mert/yasam/yazardetay/30.01.2011/1345832/default.htm

"Muhteşem Yüzyıl" a boykot tehdidi. *Farkli Haber* 8, 20 July 2011. Retrieved from http://www.farklihaber8.com/haber/guncel/muhtesem-yuzyila-boykot-tehdidi/1719.aspx

Muhteşem Yuzyil'a Muhtesem Ceza! *Haber365*, 24 April 2013, www.haber365.com/Haber/Muhtesem_Yuzyila_Muhtesem_Ceza/

MuhteşemYüzyil degil, muhtesem rezalet ve tarihe ihanet!. *diyanethaberler.com*. 10 January 2011. Retrieved from http://www.dinihaberler.com/haber/9354/-muhtesem-yuzyil--degil-muhtesem-rezalet-ve-tarihe-ihanet-.html

RTÜK'ten Aşkı Memnu'ya 'fetva' gibi tez. *Radikal*, 7 December 2012. Retrieved from http://www.radikal.com.tr/turkiye/rtukten_aski_memnuya_fetva_gibi_tez-1111053

Şeker, M. & Şimşek, F. (2011). Ötekilik Bağlamında 'Muhteşem Yüzyıl': Dizisinin Farklı İdeolojideki Gazetelerin köşe Yazılarına Yansımaları. *Turkiyat Arastirma Dergisi*, 2011. Retrieved from http://www.turkiyat.selcuk.edu.tr/pdfdergi/s29/seker.pdf

Tarihi Kriteri Kim Belirliyecek. *Radikal*, 6 December 2012. Retrieved from http://www.radikal.com.tr/hayat/tarihi_kriteri_kim_belirleyecek-1110744

The Hey Day of Turkish Content, October 2012. Retrieved from http://www.todotvnews.com/scripts/templates/despliegue_imprimir_con_foto.asp?nota=eng/Distribuci%F3n/Mercados/2012/10_Octubre/15_turquia_mipcom&numero=

Turkish TV show Behzat C draws ire of censors, adoration of public, 3 December 2012. Retrieved from http://www.pri.org/stories/arts-entertainment/turkish-tv-show-behzat-c-draws-ire-of-censors-adoration-of-public-12250.html

White, J. (2012). *Muslim Nationalism and the New Turks*, Princeton University Press.

Interviews/ Video clips

Interview with Meral Okay by Banu Guven on *ARTI* program, NTV, 6 January 2011. Retrieved from http://www.vidivodo.com/video/meral-okay-muhtesem-yuzyil/525570

News item, 'MuhteşemYüzyil gercegin tam tersi', *Samanyolu TV*, 4 January 2011. Retrieved from http://www.kure.tv/webtv/803-haber/muhtesem-yuzyil-gercegin-tam-tersi/4922-Bolum/81621/

Talk show *32. Gun*, Kanal D, 3 December 2012, episode in which *Muhteşem Yüzyil* and the controversy surrounding it are discussed. Retrieved from http://www.youtube.com/watch?v=GVvP13EtCkY

CHAPTER FIVE

NARRATIVE STRUCTURE ANALYSIS OF THE 2012 EMMY NOMINEES FOR DRAMA TV SERIES: WHAT DOES THE PILOT EPISODE REVEAL?

CÉLIA BELIM

Introduction

A TV series is a 'multimodal product, involving semiotic codes' (Bednarek, 2010, p. 17) and usually features the same characters, theme and settings (Bednarek, 2010, p. 12). The *breath of live* of TV series is narrative. Narrative is a way of comprehending space, time, and causality, and the principle by which data is converted from the frame of the *screen* into a *diegesis*—a world (Branigan, 1992, p. 36). Narrative rests on our ability to create a three-dimensional world out of a two-dimensional wash of light and dark (Branigan, 1992, p. 33).

The language conveyed in TV series—condensed in narrative, characters, stereotypes, and code among other elements—is becoming more sophisticated (e.g. Jacobsen, 2006), and some TV series persuade the spectator at a point that he becomes loyal to them, accompanying several or all seasons.

Narrative structure is about (Pointz): a) the content of a story (i.e. what a story is about); and b) the form used to narrate a story (i.e. how a story is told). These two dimensions are respectively story and plot. To analyse the story structure, 'who?', 'what?', and 'where?' questions may be used.[1]

---

[1] Examples: Where is the story set?, What event begins the story?, Who are the main characters?, What conflict(s) do they face?, What happens to the characters as they face this conflict?, Who wins the conflict?, What rewards do they receive?

"How?" questions are the preferred way to examine the plot structure. [2] According to Perrine & Arp (1983), Shapiro & Beum (1965), Williams (1986), and Zillman (1966), the following are elements of drama: 1. Plot (the sequence of events of which the story is constituted); 2. Character; Theme; 4. Points of view; 5. Symbol; and 6. Irony.

It is the main objective of this chapter to perceive, within the analysis of narrative structure (story and plot or textual analysis) and in terms of selected drama narrative analysis structures / units / elements / segments (e.g. Herman & Vervaeck, 2005; Aumont & Marie, 2009, p. 83-101), the codification of six contemporary TV series, acclaimed by both critics and public.

The *corpus* is composed of the series nominated for the 2012 Primetime Emmy Award for Outstanding Drama Series (table 1), which recognizes excellence and quality in the TV industry.[3] The pilot episode of each of the six series is chosen to analyse due to its vital functions: 1. Presentation of the story, revealing its main themes, the characters, the interactions, and other pertinent aspects; 2. Ability to convince the public, in order to grant their loyalty to the next episodes; thus the pilot episode should act, in a large sphere of competition, as a teaser or *hook* to capture TV consumers; 3. Mastery in the use of language to codify or convert the argument/the script into action: how to narrate, the plot. In VanDerWerff's words, a pilot of a TV show 'is an introduction to the characters and the world they live in. It's a statement of purpose, a way to say, "Here's what we're going to be doing." Ideally, it's also a compelling story in its own right' (2010).

In this analysis, anchored in Semiotic and Filmic frameworks, the following are privileged as techniques or methodologies:

1. *Plot or textual analysis* (Penafria, 2009, pp. 5-6; Metz, 1974, 1982, 1988): Devoted to the structuralism inspired by Linguistics, it is anchored in the division of the narrative into parts: plot analysis. Classically: the setup / beginning, evolution / development / middle, and end[4];

2. *Content analysis* (Aumont & Marie, 2009, pp. 83-86; Penafria, 2009, p. 6): This consists in summarizing the principal themes conveyed

---

[2] Examples: How is the major conflict in the story set up?, How are the main characters introduced?, How is the story moved along so that the characters must inevitably face the film's central conflict?, How is the dramatic confrontation set up as the film draws to its close?, How does the film resolve most of the major conflicts set up at the outset?

[3] It corresponds to the Academy Award (for film), the Tony Award (for theatre), and the Grammy Award (for music).

[4] Aristotle, in his *Poetics*, underlined the idea that 'A whole is what has a beginning and middle and end' (1450b27).

and codified in the TV material analysed. It should complete the sentence: 'This production is about…';

3. *The binominal 'stereotypes and clichés'* (Barthes; Joly, 2003): How TV, among another media, has been contributing to the construction of stereotypes (Walter & Baptista, 2007) and also clichés. For Barthes, the stereotype is something negative. It is 'how a society produces stereotypes, i.e., triumphs of artifice, which it then consumes as innate meanings, i.e., triumphs of nature' (Barthes, 2004, p. 33). Barthes believes that

'by virtue of certain stereotypes, a cause is expected and it is another that appears: "a woman stabs her lover": crime of passion? no, "they did not get along well in the field of politics." "An employee kidnaps the baby of her bosses": to get a ransom? no, "because she loved the child"' (Barthes, 2007, p. 62).

The stereotype is then a 'commonplace' (Barthes, 2004, p. 58), such as the femme fatale, the damsel in distress, the impractical intellectual, the mad or eccentric scientist or professor, the faithful sidekick or servant, or the tart with a heart (Bulman, 2007, p. 216). It can be seen as 'an easy and lazy way to create character' (Bulman, 207, p. 214). In turn, a cliché is

'an overused or hackneyed phrase or expression which, because of overuse, has lost its effectiveness. When we read the cliché "as black as coal" we know that it means very black but because we've heard the expression so often it is quite likely that we will just skirt over the description' (Bulman, 2007, p. 45);

4. *Deconstruction and (re)construction* (Vanoye & Goliot-Lété, 2006): This binominal is effected in two stages: (1) decomposing the movie (or TV series) into its constitutive elements. It consists in unravelling, disuniting, extracting, separating, and highlighting, and in canvassing materials that are not perceived, without fragmentation from the whole, 'to the naked eye'. (2) establishing links between these isolated elements, understanding how they combine and become accomplices to bring up a significant whole: rebuild the TV episode or fragment;

5. *Denotation and connotation*[5] (Barthes, 1989; Guiraud, 1999: 31):

[5] Denotation and the connotation are two opposite main ways of meaning. Denotation is the description provided with objectivity, factualism, and judgments of fact, and without any contagion of subjectivity and valuation. It does not allow interpretative ambiguity and variability: it only fixes one meaning. Connotation is based on value judgments, on subjectivity. In the words of Guiraud, denotation 'is constituted by the meaning objectively conceived and just like that', while connotations 'express subjective values linked to the sign, resulting from its form and its function' (1999, p. 31). For example, a uniform denotes a function and perhaps a rank and connotes a greater or lesser prestige, refinement and authority.

Denotation in order to describe the narrative elements. Connotation to discover and deepen the symbols and interpretation of some elements;

and 6. *Photogram analysis* (Penafria, 2009, pp. 7-8): The deconstruction of some photograms may reveal enriching in the purpose of extracting interpretations and finding out signs and symbols.

So, supported by these techniques, the narrative elements analysed are:

1. Plot / Form (syntagmatic or sequential elements) (Aumont & Marie, 2009, pp. 86-95; Propp, 1965; 1970). The plot analysis is mainly supported by Roland Barthes's two types of unit in a narrative (1966): functions (actions linking the story's surface) and indications (indices—static elements integrated at the thematic level);

2. Protagonist (Aumont & Marie, 2009, pp. 99-101). To characterize the protagonist, some analysis elements are used, such as: action / praxis, speech / logos, and appearance (Bednarek, 2010, pp. 18-20);

3. Themes / Topics[6];

4. Stereotypes;

5. Clichés;

6. Symbols (a symbol is an object, animate or inanimate, that stands for or points to a meaning well beyond itself);

7. Presence or Absence of Narrator;

8. Narrative Time (Chronological or Linear: follows the diachronic without inversions; Non-Chronological or Non-Linear: inverts the line of time);

9. Type of Scenario (Realist: the narrative is set in the reality of the public; Geofiction: a fictional space to embrace the narrative is created; Fantastic: another reality than the material world is conceived);

10. Time period.

---

[6] The theme is the controlling idea or central insight. It a) can be a revelation of human character; b) may be stated briefly or broadly; and c) is not the 'moral' of the story.

The option for including characters and topics in the analysis scheme is also justified by the evidence that the TV narrative often relies on these elements rather than resolving the plot's dilemmas, as happens more in a Hollywood film's narrative (Altman, quoted in Feuer, 1986, pp. 101-102).

Table 5-1. Corpus

	Boardwalk Empire	Breaking Bad	Downton Abbey	Game of Thrones	Homeland	Mad Men
Executive Producers	Terence Winter; Martin Scorsese; Mark Wahlberg; Tim Van Patten; Howard Korder; Stephen Levinson.	Vince Gilligan; Mark Johnson; Michelle MacLaren.	Julian Fellowes; Gareth Neame; Rebecca Eaton.	David Benioff; D. B. Weiss; Frank Doelger; Bernadette Caulfield; Carolyn Strauss; George R. R. Martin.	Howard Gordon; Alex Gansa; Gideon Raff; Michael Cuesta; Henry Bromell; Chip Johannessen; Meredith Stiehm; Alexander Cary; Avi Nir; Ran Telem.	Matthew Weiner; Scott Hornbacher; Andre Jacquemetton; Maria Jacquemetton; Janet Leahy.
Broadcaster (Original channel / Original run)	Home Box Office (HBO), September 19, 2010–present	American Movie Classics (AMC), January 20, 2008–present	Independent Television (ITV), 26 September 2010–present	HBO, April 17, 2011–present	Showtime, October 2, 2011–present	AMC, July 19, 2007–present

Source: Own elaboration

Boardwalk Empire

'Coward, monster, vicious brute
Friend to thief and prostitute,
Heartless, Godless, hell's delight,
Crude by day and lewd by night,
Conscience dulled by demon rum,
Liquor, thy name's
Delirium.'
(Mrs. McGarry reads this poem before Nucky Thompson's address at the Women's Temperance League)

Plot or Textual Analysis

The Form

1) **Climax**: 1920 is the year of the opening of the story narrated in the series, and Atlantic City offers the space. There can be heard the sounds of a pocket watch, birds (possibly seagulls), a bell, and the motor of a boat approaching. The first scene is coincident with the sounds mentioned and consists of images of a group of bootleggers, on a night with little moonlight—indicating the obscurity of their activity—, smuggling Canadian Club whisky from boats into the United States. Later, these men are ambushed in their truck and held at gunpoint by two men in ski masks.

2) **Flashback**: Characters' presentation and interactions: Then, the chronological line is cut and, in a flashback ('three nights earlier'), the spectators are invited to meet the characters. The protagonist, Enoch Malachi 'Nucky' Thompson, the Atlantic County treasurer, appears delivering the keynote address at a Women's Temperance League rally on the eve of Prohibition. Nucky reports a horrifying episode from his childhood: he was abandoned by his father during the Great Blizzard of 1888 and captured three wharf rats in order to feed his family. This heart-rending tale proves to be a lie: a politician says the words that the crowd wants to hear—says Nucky to Jimmy Darmody, his driver. His rhetorical abilities inspire the all-female audience, especially one lady who deserves the focus of the camera: Margaret Schroeder.

Nucky leaves the rally early and promptly heads to Babette's Supper Club. Here, a group of elected officials, including Atlantic City's mayor and Nucky's brother, Sheriff Eli Thompson, are celebrating the onset of Prohibition and the potential lucrative bootlegging that it represents. Nucky, taking advantage of this social environment, introduces Darmody, who has just returned after serving in World War I. At midnight, the party reaches its zenith, with partygoers acclaiming the 'death' of alcohol. The next morning, Darmody and Angela, his wife, talk about the continuing of his studies at Princeton, but he shows a preference for continuing his work for Nucky. The unknown lady takes the initiative to go to Nucky's house, displaying an advanced pregnancy, and asks a job for her husband, Hans Schroeder.

When the night falls, Nucky and Darmody visit Mickey Doyle's funeral parlour, a disguise for an alcohol distillery.

In a subsequent scene, Nucky dines with four major mob bosses, New York's Arnold Rothstein and Lucky Luciano and Chicago's Big Jim Colosimo and Johnny Torrio, who agree to start buying Nucky's seaborne liquor shipments. Nucky agrees to sell his next shipment to Rothstein, specifiying the condition that the buyer's own men pick it up. While Darmody waits outside for Nucky, he befriends Torrio's driver, Al Capone. Meanwhile, Margaret is assaulted by her husband in a way that suggests routine behaviour.

3) **Teaser**: end of the flashback, continuing of the narrative, and glimpse of the Nucky's future or path: on the day of the heist, Darmody recruits Capone to hijack Rothstein's whiskey shipment and the episode returns to the beginning. Nucky deduces that the robbery's perpetrator is Darmody. Meanwhile, Nucky finds out about Margaret's hospitalization. The offender (Darmody) confirms the suspicion, counting on Nucky's forgiveness and asking for his intervention with more aggressive criminal enterprises, claiming that war has left him with no other future than violence.

Nucky shows curiosity about his destiny. He peers into a fortune teller's shop, watching her read someone's palm. The camera frames him through the glass front door. The writing on the door, standing out in front of his face, asks 'What does the future hold for you?' This question seems to be directed as much at the audience as it is at Nucky himself.

The narrative continues with Eli Thompson and his deputies kidnapping and beating Hans to death on the high seas. These scenes are intersected with ones that show the assassination of mob boss Big Jim Colosimo in his Chicago restaurant.

The episode ends with Nucky visiting and delivering flowers to a hospitalized and widowed Margaret. This final scene alludes to a possible romance between the two of them.

These final scenes possibly embody Nucky's dual and antithetical facets: wickedness and candour.

Story Analysis

Protagonist

Enoch 'Nucky' Thompson is a successful, charming, and intelligent man. He is a widower and has no children. Despite publicly supporting the banning of alcohol, he conspires with local politicians to profit from bootlegging schemes. In terms of personality, he exhibits a Manichean dimension, mixing his involvement in illegal and criminal activities with a humane and sensitive part, shown in his delivery of sums of money to Margaret and, on another occasion, to Darmody. 'Nucky is one of numerous "Boardwalk Empire" characters with two selves, public and private' (Seitz, 2010).

He reveals a sense of humour. For example, in the scene Margaret looks for a job in Nucky's home for her husband and, after the treasurer hands her a wad of cash, she accepts and says she will 'be honoured' to name her child after him. Nucky answers: 'Enoch? You couldn't possibly be so cruel.'

Themes, Stereotypes, and Clichés

a) A gangster ambience, evidencing a 'savage law' orientation, which it is an irony due to the label of the United States as government 'of the people, by the people, for the people' (Lincoln, in Richards, 1992, p. 27).

b) The clandestine manufacture and smuggling of drink (bootlegging), and the casino ambience, gaming, extravagance, and vice, in the context of Prohibition or 'the noble experiment'.[1]

c) The composition of gangs that defend their interests by the use of guns and violence. Violence seems to be the best solution to a problem. This is evident mainly in the two assassinations at the end of the pilot episode.

d) Subservience of women, on one hand as objects of desire and pleasure (dancers), and on the other hand as executors of domestic and maternal tasks.

e) Maltreatment of women derived from alcohol and vices. This is Margaret's case, in which her pregnancy does not inhibit her husband from beating her.

f) A certain respect for women, present in the conservation and exhibition of the portrait of Nucky's wife and in the involvement of ladies in a league.

[1] The expression was coined by President Herbert Hoover, who wrote to an Idaho senator in 1928: 'Our country has deliberately undertaken a great social and economic experiment, noble in motive and far-reaching in purpose.'

g) Hypocrisy: Nucky discourses after a previous camera focus on a poster that shows the maxim: 'Lips that touch liquor shall never touch mine' and, by the sequence of narrative, we perceive that he is the head of illegal enterprises linked to alcohol.

h) Curiosity about the future, evidenced in Nucky's proximity to the shop of a palm reader, the camera framing of the treasurer's face behind a glass door, and the meeting of eyes between the two of them.

i) 'First rule of politics, kiddo: never let the truth get in the way of a good story.' (Nucky Thompson)

j) 'Do you know who dies for their country? Fucking rubes.' (Nucky Thompson)

l) 'You young fellas... no appreciation for the art of conversation.' (Nucky Thompson)

Symbols

a) The scene in which Nucky looks thoughtfully towards the sea, is filled with poetic pigments and can be representative of a sort of premonition of the destiny or path conceived by the producers for Nucky at the outset of the narrative: alone, with a long bench to seat people, possibly his partners or family members that accompany him during his life's path—the bench is contiguous to him—, and with a entire sea and sky to conquer. The green colour of the bench can symbolize annual renewal of nature, hope, longevity and immortality (Lexicon, 1990, p. 203).

b) In the screen, Nucky approaches the vitrine and peers at the seer, manifesting the curiosity of the treasurer about his future.

Breaking Bad

Jesse: *[about Walter's meth] This is glass grade. I mean, you got... Jesus, you got crystals in here 2 inches, 3 inches long. This is pure glass. You're a damn artist! This is art, Mr. White!*

Walter: *Actually it's just basic chemistry, but thank you, Jesse. I'm glad it's acceptable.*

Jesse: *Acceptable? You're the goddamn Iron Chef! Every jibbhead from here to Timbuktu is going to want a taste.*

Plot or Textual Analysis

The Form

1) **Climax**: Some images of landscapes, static, unpopulated by human figures landscapes, and devoid of sound are the substance to the initiation of the pilot episode. Sound and movement only appear when a pair of trousers, seeming to fall from the sky (it falls from the upper edge of the screen) lands on the ground, and then immediately a Winnebago, in a disarticulated direction, runs over it. The driver wears a gas mask, as does the man that occupies the passenger seat. This man is unconscious and possibly dead. The vehicle traces a chaotic and speedy route in the dust road and, in this disorientation, two bodies slide across the Recreational Vehicle (RV) floor until the RV veers into a ditch and stops. Suddenly, the driver leaves the van; he is wearing nothing but underwear and the gas mask already mentioned. He comes inside the RV to get a video camera, wallet, and gun. He records an enigmatic goodbye message to his wife and son, while, in a characteristic sound-scenario of the pursuit of bandits, a noise suggestive of police sirens echoes.

2) **Flashback**: Characters' presentation and interactions

2.1) The story enters a flashback ('three weeks earlier'), exactly on the 50th birthday of the protagonist, Walter White. At breakfast, we meet his wife, Skyler White, and their disabled son, Junior. His wife prepares him an original breakfast: a plate of eggs topped by veggie bacon spelling '50'. Afterwards, Walter drops off Junior at the local high school where he is an ineffectual chemistry teacher. In the class, students show lack of attention to Walter's lessons.

Later, Walter, who collects additional income at a car wash, is witnessed by one of his more disrespectful students, who laughs and takes photos while Walter is wiping down the tires of his car. Publicly humiliated, he returns home, where a surprise birthday party is prepared for him. Walter's brother-in-law, Drug Enforcement Administration (DEA) Agent Hank Schrader, present in the party, appears in a news report documenting his bust of a local methamphetamine lab. It seems Walter is curious due to his question on how much money was recovered at the crime scene. In the kitchen, during a female conversation, the spectator obtains the information that Skyler is pregnant.

The next day at the car wash, Walter collapses and is given medical assistance. The diagnosis is the worst: a developed inoperable lung cancer, leaving him little time to live. Returning home, he does not share this worrying diagnosis with his wife.

Walter psychologically explodes at work in the car wash, verbally beating his boss. Then, Walter calls Hank and asks to accompany him in a drug bust. While Walter waits in the squad car, he sees his ex-student Jesse Pinkman, 'Cap'n Cook', escaping through a window. Later, Walter proposes to Jesse that they be partners in methamphetamine activities.

Walter and Jesse dialogue to decide the operationalization of their partnership. Jesse proposes to buy a used Winnebago RV, so that this way they can have a mobile base, harder to find by the authorities. Walter, that day, seems to embody a new personality, once he defends his son against a group of male bullies, to the surprise of his family.

2.2) Walter and Jesse drive to the desert. Before some static images can be seen, recalling the ones at the start of the episode, trying to complete the time gap. Walter strips down to his skivvies and produces the purest crystal meth Jesse has ever seen, as he says.

Jesse takes a sample to Krazy-8, a drug dealer. But he distrusts the loyalty of Jesse to Emilio—who appears—and Jesse leads them to the desert, where Walter is. Emilio recognizes Walter, and the two dealers try to kill the teacher. Jesse becomes unconscious in the fights. Walter, trying to save his life, promises to teach them his formula. The traffickers decide to save his life.

In the Winnebago, while Walter prepares ingredients, Emilio is smoking and Walter asks him to put out the cigarette. Emilio throws the cigarette outside and it ignites a brush fire. Meanwhile, Walter uses a firer explosion to escape and shuts the two traffickers in the vehicle, getting both unconscious. Some bullets fly, but Walter has energy to keep the door closed.

3) **Teaser**: End of the flashback, continuing of the narrative, and evidence of a *new* Walter: Confronted by an exterior fire generated by the discarded cigarette, Walter has to leave that place. So, he unties Jesse, shoves a gas mask on him, and puts him in the passenger seat. The scene reinitiates the scene of the beginning. Walter stops the vehicle and stands in the middle of the road, listening the sound of sirens that he associates with the police. While he waits for the cops' arrival, he tries to kill himself, but, to his dismay, the gun's safety catch is activated. He fights with the gun, but, in this *battle*, he sees fire engines, and not police cars. So, he escapes an eventual capture.

At home, he shows a new sexual vitality to his wife, which makes her ask him: 'Walt, is that you?'" There is the insinuation that a depressed, frustrated, and repressed Walter will be replaced by a new and regenerated one.

Story Analysis

Protagonist

Walter White is a man of fifty. He embodies a weak personality with little vitality and it seems he carries *all the world's weight*. He is married and has a son with deficiencies. He is a teacher of chemistry, but students do not pay attention to his teaching. So he does not feel professionally fulfilled. He has a part-time job in a car wash to get more money in order to balance the familiar budget. He discovers that he has lung cancer and engenders an audacious ploy—as it involves illegal activities—to change his life.

Vince Gilligan, the producer, has confided it is hard to write for Walter White because his character is so dark and morally questionable: 'I'm going to miss the show when it's over, but on some level, it'll be a relief to not have Walt in my head anymore' (Giligan in Segal, 2011).

'Physically, to create Walter White, I use my dad (…)' 'My dad is 87 years old. I'm not going to dodder, but Walter is always a little hunched over, never erect. The message to the audience is that the weight of the world is on this man's shoulders' (Giligan in Segal, 2011).

About the reasons for the turnaround in his life,

'*Jesse Pinkman says*: Wait wait hold on. Tell me why you are doing this... seriously.

Walter H. White: Why do you do it?

Jesse Pinkman: Money... mainly.

Walter H. White: There you go.

Jesse Pinkman: Nah come on... man, some straight like you giant stick up his ass all of a sudden at age what 60 he's just going to break bad.

Walter H. White: Fifty.

Jesse Pinkman: It's weird is all okay it... it doesn't compute. Listen if you've gone crazy or something. I mean if you've if–if–if you've gone crazy or depressed I'm... I'm just saying that that's something I need to know about. Okay, I mean that that affects me.

Walter H. White: **I am... awake.'** (IMDb, 2008).

During the episode, his wife comments that he is quieter than normal. But, at the end, surprised at his sexual vigour, she asks: 'Walter, is that you?'

Krazy-8 labels him as 'artist', confirming his ability and mastery at producing the 'crystals.'

Themes, Stereotypes, and Clichés

a) A man that displays a boring life, but reaches a limit and injects a turnaround in his life, practicing illegal activities with his chemistry training.

b) Mental linkage between Mexicans and illegal drug activities made by DEA agents.

c) About confection of methamphetamines: 'This ain't chemistry—this is art.' (Jesse)

d) 'You're a drug dealer... negotiate.' (Walter to Jesse)

e) An apparent happy family.

f) The classical antithesis between the police and the bandits.

g) The protagonist does not practice moral activities, but, in certain way, is admired due to the way he inverts the failed and erased life he has (anti-hero).

Symbols

a) The phosphorus, that he lights up before (or during) the decision to embark on an illicit life, can benefit the symbolism of fire as something sacred, purifying and regenerating (Lexicon, 1990, p. 99).

b) Iconic and plot narrative (static landscape images devoid of sound) parallelism with the beginning of Stanley's Kubrick's movie, *2001: A Spacy Odyssey* (1968). The bone, falling—with a blue sky in a backdrop—and composing an analogy with the spacecraft, shows the colossal change of man from the age of primates to a futuristic age. Walter also experiences a colossal change in his life. The pants take the place of the bone.

Downton Abbey

'Every mountain is unclimbable until someone climbs it, so every ship is unsinkable until it sinks.'
(Lord Grantham)

Plot or Textual Analysis

The Form

The opening titles merge with the start of the episode, in an exhibition and tribute to the technologies of the time period narrated. First, a close up shows the accuracy of a hand tapping out Morse code. Then a train in movement, circulating between landscapes, fills the screen. Through the glass of the train's widow, John Bates is seen in the train, contemplating the scenery. The noise of the train and the music of title sequence are heard simultaneously. The sounds of the music, the train, and the telegraph mix, but can be identified. Then some electric cables conduct or invite the viewer's eye to a workspace, where a lady receives a message. The lady expresses admiration and shock, confirmed by her words: 'Oh, my God!' The man, her colleague, says: 'That is impossible!' This state of mind feeds curiosity about the veiled issue.

1) **Awakening and presentation of the servants**: The effective episode begins with a close up of a telegraph functioning, giving passage to the building (Downton Abbey) where the action takes place. The date is also revealed: April 1912. Inside the building, we hear the bell for awakening, and then the agitation of servants—who are *parading* in order to be introduced to the public—, implying preparation for breakfast.

The next ringing symbolizes the awakening of the noble family that lives in Downton Abbey. The paperboy's arrival is also announced by a bell.

2) **Presentation of Grantham's family and their interactions**: Robert Crawley, the Earl of Grantham, comes down the stairs accompanied by his dog and, already in the breakfast room, talks with Charles 'Charlie' Carson, the butler, about the news about the sinking of the Royal Mail Ship (RMS) Titanic in April 1912. Then he prepares to take breakfast with his daughters. But, early in the meal, he receives a telegram from George Murray, reporting the presence of his relatives James and Patrick in the passenger list of the Titanic and their hypothetical fatal destiny.

His distress is higher due to the fact that the last kinsman mentioned, the son of the then heir presumptive, is the fiancé of Lady Mary, Lord Grantham's eldest daughter, thus affecting the succession of the family's estate. As the Grantham couple have no male heirs and a device stipulates that the fortune—the (fictional) Earldom of Grantham, which endows both title and estate—should be in the hold of the male, annihilating the inheritance of the bride, there is a problem to solve.

The Earldom of Grantham receives a new worker: John Bates, to the unoccupied function of Lord Grantham's valet. His recruitment causes some agitation and strangeness, as he has a locomotive deficiency and the building has long staircases.

The images, putting Bates in the centre of the stairs in a high camera angle that suggests the character's being crushed, considering his physical-motor deficiency, allude to the great challenge he will face.

Receiving the news of her fiancé, Lady Mary shows little interest in wearing mourning, converging on the idea of her meagre and scarce feelings for the promised husband.

3) **Teaser**: Continuing of the aim of finding a bachelor relative for Lady Mary: In a breakfast atmosphere, the episode finishes with the receiving of a letter from a young man. A distant male cousin comes into play, Matthew Crawley, a middle class lawyer from Manchester, as heir presumptive to the countess's fortune. The public can envision the contentions that will follow.

The narrative ends as it begins: with the receiving of a message.

Story Analysis

Protagonist

Lord Grantham, the patriarch of Downton Abbey, embodies traditional aristocratic values. He is married to Cora Crawley and they have three daughters.His major challenge is to maintain the Grantham Earldom's fortune within his family. Thus with a view to this objective he endeavours to find a relative as bridegroom for his oldest daughter. It is only by this path, that he can avoid the family fortune's falling into extra-family hands.

The Earl served in the Second Boer War between 1899 and 1902, where he met John Bates, at the time his batman.

He first appears in the series descending the imposing stairs of Downton Abbey accompanied by a dog, symbol of fidelity. The wisdom attributed to the dog makes him the ancestor of civilization and the bearer of fire to men (Lexicon, 1990, p. 44). A white dog generally signifies goodness and devotion to the person at whose feet he is represented (Lexicon, 1990, p. 44).

These qualities (fidelity, wisdom, goodness, devotion) can be exported to Lord Grantham's personality as he is a loyal man. He is very protective of his family and servants, and, in many cases, treats the servants almost like family members. He also reveals sensitivity and humanity as he gives a professional opportunity to the disabled Bates and does not let him go after a decision in that sense. He reveals 'a kindness that is the root of his strength. He is fiercely protective of both his servants and his family' (Daniel, 2013).

Themes, Stereotypes, and Clichés

a) Intra-group marriages and marriages of convenience, evident in the search for a bachelor with kinship ties for Lady Mary in order to retain the fortune within the family. In the case of the Grantham couple, the Earl has fallen in love with the Countess.

b) Classical conflicting relationships between 1. mother-in-law, Violet

Crawley, the Dowager Countess of Grantham, and daughter-in-law, Cora Crawley, the Countess of Grantham; 2. sisters: Lady Mary and Lady Edith; 3. servants / colleagues, mainly Sarah O'Brien, Lady Grantham's lady's maid, and Thomas Barrow, the First Footman, against Bates, Lord Grantham's valet.

c) Social differentiation, visible in the segregation from servants, revealing higher status. Despite the friendly relationship that transpires between the Crawley family and the servants, there is an obvious segregation notorious at meals.

Lord Grantham's first dialogue in the series confirms this theme: 'I understand most of the ladies were taken off in time,' says Carson. 'You mean the ladies in first class,' says Lord Grantham, with furrowed brow. 'God help the poor devils below decks on their way to a better life.'

d) Professionalism, patent in Carson's constant worries tending to perfectionism.

e) 'Nothing in life is sure' (Beryl Patmore, the cook).

f) Homosexuality displayed in the ended romance of Thomas and the Duke of Crowborough.

g) Mediocrity of the human spirit, played by Sarah O'Brien and Thomas Barrow, who seek voluntarily to prejudice Bates. They constantly hatch schemes, intending to remove Bates from service at Downton. Thomas Barrow also tries to blackmail his former lover, the Duke of Crowborough.

h) Beauty of spirit and feelings, patent, for example, in the relation between Anna Smith and Bates, in paternal and maternal feelings, and in Lord Grantham's complacency, compassion, and pity for Bates.

i) Humanity in men, mainly present in Lord Grantham. Also Bates shows this quality: when confronted with the order to abandon the Earldom, he is seen crying by Anna Smith, the head housemaid.

j) References to technological progress, such as the telegraph, electricity, and the train, charactering the industrial era and paying tribute to a technological advent.

l) The sinking of the Titanic, which sparked a panoply of events, specifically in a micro-context: Downton Abbey.

Symbols

a) The newspaper and telegrams are symbolic of news. The telegram is symbol of technology's progress, ease of communication, and the announcement of news with an impact.

b) The telegraph symbolizes the progress of technology in the time narrated.

c) The bell suggests a daily ritual, a starting point (e.g. the outset of a journey), a warning of something (e.g. the arrival of the paperboy).

d) The alternative syntagm in the opening hints that the coming of the paperboy bringing the news has an association with the lady (Lady Mary) who is peering through the window. This impression will be confirmed later.

e) The dog, symbol of fidelity[2].

f) Mourning signifies culturally patterned expectations about the expression of grief (Hooyman & Kiyak, 2008, p. 569). The ritual of mourning recognizes the loss (Botting, 2002, p. 92), fills a hole in the reality (Botting, 2002, p. 93), and is a demonstrator of sorrow at the lack of someone who is esteemed.

g) Electricity compared to a diabolic and appalling activity. Vestiges of superstition and ignorance. References to the future surrender to technology.

Anna: Why didn't you put the lights on?

Daisy: I daren't.

Gwen: **Well, it's electricity, not the devil's handiwork**. You'll have to get used to it sooner or later.

Game of Thrones

'Winter is coming'

'I know I broke my oath. I know I'm a deserter. I should've gone back to the Wall and warned them, but... I saw what I saw. I saw the White Walkers. People need to know. If you can get word to my family... tell 'em I'm no coward. Tell 'em I'm sorry.'

(Will—the ranger that escaped—to Lord Ned, just before his execution)

Plot or Textual Analysis

The Form

1) **Beyond the Wall**: Three rangers of the Night's Watch scout beyond the Wall, a magnificent barrier of ice that signals the north end of the kingdom. These rangers are surprised by supernatural creatures and undead wildlings that kill two of them.

2) **The presentation of the characters and their interactions**:

2.1) **In the North**: In the sequence of finding mutilated bodies and watching the death of his two fellows, the survivor ranger, Will, deserts

---

[2] See in detail the symbology of the dog, and particularly the white dog, in the presentation of the protagonist Lord Grantham.

the Night's Watch. But he is captured. Then in the Winterfell, the Lord Eddard 'Ned' Stark and his family (his wife, Lady Catelyn Stark, their five children (Robb, Sansa, Arya, Bran, and Rickon), and the illegitimate son, Jon Snow) are introduced and Lord Ned is convened to the execution of the deserter, according to ancestral tradition. Upon the male Starks' return, they find a dead stag, sigil (seal) of House Baratheon. Then, they find a dead direwolf (sigil of the Starks) and her surviving pups, and take them.

2.2) **At King's Landing**: The sound of bells reveals a death. Queen Cersei and her twin brother, Jaime Lannister, are observing the dead body of the Hand of the King, Jon Arryn. The scenario of suspicion has been formed once they discuss whether he revealed (or not) any dangerous information about the two siblings before dying.

2.3) **Back at Winterfell**: Catleyn receives a letter, brought by a raven, announcing the death of Lord Arryn. An additional message reports that King Robert Baratheon's court (the King himself; his wife, Queen Cersei; their three children; and Cersei's brothers, Jaime Lannister and Tyrion Lannister, a dwarf) is coming to Winterfell. The King decides to name Lord Ned as the new Hand of the King. At night, Catelyn gets a message from her sister, Lord Arryn's widow, revealing her suspicion that the king's in-laws, the Lannisters, murdered her husband. Lord Ned accepts the invitation of the King to be the next Hand of the King.

2.4) **Pentos, Across the Narrow Sea**: The exiled prince Viserys Targaryen plots to overthrow King Robert in order to reclaim his father's throne. To achieve this aim, he arranges the marriage between his sister, Daenerys, and a stronger warlord, Khal Drogo, the leader of a nomadic horde of Dothraki.

3) **Teaser**: Bran, the youngest son of the Stark couple, has a particular ability to climb on the walls. In a climbing, he stumbles on Queen Cersei and Jaime having sex. In order to keep the incestuous relationship a secret, Jaime shoves Bran out of a high window.

Story Analysis

Protagonist

Lord Ned Stark is the head of House Stark and Lord Paramount of the North. He is devoted to his family, a loyal friend, and a noble and honourable lord. He accepts the invitation of the King to be the Hand of the King. He fulfils ancestral traditions.

Themes, Stereotypes, and Clichés

a) Strength of ancestral traditions, patent in the decapitation of the survivor ranger;

b) Mystery / enigma: the appearance of the White Walkers (the demonic creatures who ambushed the rangers), which have been considered extinct for a thousand years;

c) Suggestion of fights, wars, alliances, and obstinacy (prince Viserys Targaryen, for example) to seize power;

d) Suspicion of the murder of Lord Arryn;

e) Incest between Queen Cersei and her twin brother, Jaime Lannister;

f) Arrangement of marriages in order to serve interests;

g) 'The man who passes the sentence should swing the sword.' (Eddard Stark);

h) The discriminated: the"bastard", Jon Snow and the dwarf, Tyrion Lannister;

i) Complementarity or Manichaeism between intellectual and physical powers: 'My brother, Jaime, has his swords and I have my mind. And a mind needs books like sword needs a whetstone' (Tyrion Lannister to Jon);

j) 'A Dothraki wedding without at least three deaths is considered a dull affair.' (Magister Illyrio Mopatis).

Symbols

a) The number three (three rangers): three is a sacred number, according to ancient sages. It is considered a symbolic and mysterious number; at one time, among the Greeks and Romans, it had an occult power, inspiring enigmas, superstitions, beliefs, proverbs, and popular sayings (Mendes, 2006, p. 24). As the first odd number, it is an active number, the number of heaven (Julien, 2012, p. 440) and, for the Chinese, it is a perfect number (Mendes, 2006, p. 25). According to the *Book of Rituals (Li-ji)*, man, as the intermediary between heaven and earth, corresponds to the number three.

b) Arya Stark hits the target unerringly with an arrow, manifesting a warrior spirit and skill.

c) Hand of the King: 'Their days are too long, and their lives are too short.' (Jaime Lannister)

d) The raven as messenger: A figure of foreboding, harbinger of disease, war, and death. The Bible includes the raven among the most impure animals (Lexicon, 1990, p. 67).

e) A dead stag, sigil (seal) of House Baratheon: Idea of a presage. 'As an embodiment of the male principle in nature, the stag is frequently paired with great goddess imagery in myth and art' (Werness, 2006, p. 390). The negative dimension of stag symbolism in Christianity stems from the animal's connection with male sexuality. In this association, the stag, like other horned animals, becomes a symbol of the devil and demons (Werness, 2006, p. 391).

f) A dead direwolf (sigil of the Starks) and her surviving pups. Curiously, the survivors are as many as the Stark children, even an albino runt for Jon. The Stark family take the animals. The bear's magic

'came from his ability to wander between life and death. In the winter it descended into a cold grave in the ground and in the spring it came out, alive and well. Understandably, the bear became a symbol of healing and the cycles of life and death, or resurrection.' (Swenson, 2000, p. 15).

In China, the bear 'is a male symbol of strength and bravery, heralding the birth of boys, in harmony with home, and mountain' (Werness, 2006, p. 36).

g) Bells announce death and its ceremonial.

h) The statue of Lord Ned's sister, killed by Targaryens. The impression that she was loved by the King, when he says: 'She belonged with me.'

i) Dothraki folklore: 'Do you see how long his hair is? When Dothraki are defeated in combat they cut of their braid so the whole world can see their shame. Khal Drogo has never been defeated... He's a savage of course but he is one of the finest killers alive.' (Viserys Targaryen)

j) The two most emblematic wedding gifts for Daenerys:

j1. 'Songs and histories from the Seven Kingdoms' (Jorah Andahli), evoking the grandiosity of the Seven Kingdoms.

j2. 'Dragon's eggs, from the Shadow Lands beyond Asshai,' said Magister Illyrio. 'The eons have turned them to stone, yet still they burn bright with beauty.' These stone eggs can symbolize the legacies of the Targaryen dynasty.

l) The symbolism of a bastard:

'*Tyrion Lannister*: Let me tell you something, Bastard. Never forget what you are, the rest of the world will not. Wear it like armour and it can never be used to hurt you.

Jon Snow: What the hell do you know about being a bastard?

Tyrion Lannister: All dwarfs are bastards in their father's eyes.' (IMDb, 2011)

m) Illegitimate son: Jon's presence at Winterfell is a source of friction between Lord Ned and his wife.

Homeland

Saul: *What were his exact words, please?*
Carrie: *An American prisoner of war has been turned.*
Saul: *He said this in English?*
Carrie: *Yes, he whispered it into my ear right before the guards pulled me away.*
Saul: *Why would he use the expression 'turned'?*
Carrie: *He mentioned turned, working for Abu Nazir.*
Saul: *Why am I just hearing about this now?*
Carrie: *Because until ten minutes ago, I did't know there were any POWs alive in Iraq or Afghanistan.*
Saul: *So you're suggesting that Abu Nazir planted intelligence on his own safehouse just so we could recover Sergeant Brody?*
Carrie: *I realize it sounds like a reach.*
Saul: *To say the least. Why not just drop him near a checkpoint and make it look like he escaped? Why would you sacrifice thirteen trained fighters?*
Carrie: *Because Abu Nazir is playing the long game. This way no one suspects a thing.*
Saul: *Except you.*
Carrie: *Yeah, except me!*

Plot or Textual Analysis

The Form

1) **Presentation of the co-protagonist**: The episode starts in Baghdad with a Homeland Security agent, Carrie Mathison, who enters a prison, in order to collect information from a prisoner, a bomb maker who is soon to be executed. He whispers to her crucial information relating to an Abu Nazir (the leader of al-Qaeda)'s attack: an American prisoner of war has been turned (into a traitor) (we will know this later in the episode).

2) **Presentation of characters and their interactions**: Ten months later, Carrie is at her apartment. A triangle of pictures can be observed, in which Abu Nazir is at the top. Carrie, in a bustle, occupies herself with her hygiene, takes a pill and leaves home. She is driving to the Central Intelligence Agency (CIA) offices, to participate in a meeting at which she is late. The meeting's chair, the Director of Counterterrorism David Estes, is briefing on an overseas mission that has rescued a US prisoner of war, Marine Sergeant Nicholas Brody. He has been missing, presumed dead since 2003 (eight years ago). Carrie asks her chief about the prisoner's partner, Thomas Walker, receiving the answer that according to Brody he is dead. Outside, she comments to Saul Berenson about the content of the

Baghdad prisoner's whisper.

Brody is returning to the United States of America in 22 hours, and Carrie decides to monitor Brody's house with cameras and wiretaps. She explains this surveillance to Saul: 'We need eyes and ears on Brody from the minute he steps off that plane.' In a subsequent scene, Brody, looking more like a caveman, is getting all cleaned up. He calls his wife, Jessica, who is having sex with another man. Jessica tells the surprising news to her children, Dana and Chris.

In a plane, Brody has been told that he will have some time with his family and then he will meet the vice-president and the media. Nervous, Brody vomits, but David Estes, on the ground, says he had better be ready, waving as a hero like it is the Macy's Day Parade.

Brody encounters his family and thanks the Special Forces for saving his life. At the same time, their house is being prepared for surveillance. Brody runs into Mike Faber, his best friend and a Marine fellow. The spectator recognizes the man as the one who was having sexual relations with his wife, implying, in the story, a love triangle.

In another scene, Brody is debriefed. Carrie puts some questions to him. One on the reason he was kept alive for eight years. He denies knowing Abu Nazir. This negation is unmasked by a flashback of Brody that shows he is hanging naked in a room and a man approaches and gives him water.

Outside the offices, Brody goes to Bluemont Park and lies to his wife about what he will do, increasing inklings in Carrie, Virgil, and Max, who are monitoring him. But the doubts fade once he meets Walker's wife. Brody tells her that Walker was beaten to death and he was not watching. However, the flashbacks revealing someone punching his colleague / friend and him standing there contradict his words. Afterwards, Brody goes to a family barbecue and possibly distrusts the proximity between Jessica and Mike.

Saul finds all the surveillance equipment in Carrie's house and scolds her. There is the impression of Carrie trying to seduce him but he refuses her advances.

Carrie is at a bar watching Brody on the news and observes his fingers moving, suspecting that these movements could be a code, a pattern, possibly intended for a handler or sleeper cell. She comments about this suspicions to Saul, but he says it is nerves. But Carrie insists and shows that, in television images, he moves his right hand fingers the same way. Then, Saul, convinced, says it is a matter to be investigated by cryptographers.

3) **Teaser**: Suspicion about Brody's intentions: At Brody's house,

Jessica wakes up and Brody is not there. He is running through the deserted streets of Capitol Hill. This not a light jog as he is really pushing himself up a long steep climb (Gansa, Gordon & Raff, 2010, p. 60). Simultaneously, he is having flashbacks of himself, under Abu Nazir's command, repeatedly punching Walker in the face. These flashbacks are overlaid with memories of conversations with Carrie and Walker's wife. In these memories, Brody starts crying and Abu Nazir puts Brody's head on his shoulder to comfort him, indicating an intimate relationship or the initiation of one. In front of the Capitol, Brody stops running and stares at the building.

Story Analysis

Protagonists

Carrie Mathison

She is a Homeland Security agent. She is brave, obstinate, stubborn, fearless, and, at the start of the episode, she disobeys the order of her chief, David Estes, to finish the Baghdad operation. Carrie is 'incredibly bright, at times dangerously bright, and formidable and focused, even compulsive, myopic (...) She's also very, very sensitive and vulnerable' (Danes, in Levin, 2011).

Pursuing her feeling of suspicion about Brody's real intentions, Carrie violates twelve federal laws to monitor Brody's behaviour and words. Damian Lewis explains 'She's like a sniffer dog; she can smell when something's not right, she's just got to follow the trail' (in Levin, 2011).

The pills she ingests launch the suspicion of disease or disturbance.

She has a close relationship with Saul, who assumes a protective role towards her.

Nicholas Brody

A Marine Sergeant captured in 2003 by al-Qaeda in Iraq, Middle East, and presumed dead in the USA. He is rescued by Delta Force after eight years of captivity. His body is scarred by the torture inflicted on him. After his rescue and return home, he is hailed as a war hero.

His flashbacks reveal that his words convey lies, feeding the suspicion that he is not a real hero, but can be a possible villain, who is engendering a terrorist attack on American soil, due to his proximity to and plans with Abu Nazir.

He is married and a father.

Themes, Stereotypes, and Clichés

a) National Security, looked after by the CIA, and patriotism;

b) Obstinacy and persistence, played mainly by Carrie;

c) The stereotype of a hero: 'And I need him smiling and waving like a hero, like it's the fucking Macy's Day Parade' (David Estes's quote); 'The return of a hero' (TV headline);

d) Heroism *v.* anti-heroism;

d) Protectiveness of Saul towards Carrie;

e) Concepts of family and friendship;

f) Love triangle: Brody, Jessica, and Mike;

g) Consequences of war conflicts;

h) Terrorism and suspicions;

i) The stereotype that the USA's enemy is Islamic (Abu Nazir);

j) Construction of lies;

l) Mediatization.

Symbols

a) Baghdad with typical Islamic buildings (dome of the mosque and minaret) and clothing (the masculine taqiyah and feminine hijab);

b) Washington, with its magnificent and historical icons:the obelisk Washington Monument, the Jefferson Memorial, and the White House;

c) Abu Nazir at the top of a pyramid, indicating he is the presumed head of a conspiracy against the USA;

d) The pill that Carrie takes, alluding to a possible disease or disturbance;

e) Image of Brody's rescue, characterizing him as a caveman and showing his capture in an underworld, where anything might be done to survive. It is Darwin's general law leading to the advancement of all organic beings (2010, 197): 'the strongest live and the weakest die.' This assumption can determine two interpretations of Brody: 1. He can be seen as a hero, as he survives; 2. He can be suspected as he does everything to survive, not looking at means to ends (Machiavellian principle, 2007);

f) Brody's pause and contemplation of the Capitol: hint at a possible duel between this man and his country, which is symbolized in this scene by the legislative centre of the US Government.

Mad Men

'Smoke Gets in Your Eyes'

Plot or Textual Analysis

The Form

'Mad Men a term coined in the late 1950's to describe the advertising executives of Madison Avenue. They coined it.'

(Before the entrance to the episode)

'BUSBOY: Finished, sir?
DON: Yeah. Got a light?
The busboy pulls out a pack of matches from the back of his "Old Gold's" and lights Don's cigarette
Ah, an "Old Gold" man.
(inhaling)
"Lucky Strike", here.'
(Initial dialogue, Weiner, 2006, pp. 1-2)

1) **Presentation of the protagonist**: In a room packed full of people enjoying life and smoking, there is a man sitting in a corner. The camera shows the protagonist, Don Draper—the creative director for Sterling Cooper, an advertising agency—, from behind, maintaining suspense briefly about his identity. 'The opening sequence of season one shows Don, doing what any worthy modern marketer should do, taking inspiration from a brand advocate and conducting a piece of consumer research (albeit from a minute sample of one)' (Stott, 2003).

2) **One night and one day in the protagonist's life**: After the scene in the bar, which introduces and confirms the thematic scenario of the episode and series, Dan knocks on the door of a woman, the artist Midge Daniels, with whom he spends the night, discusses his following meeting with Lucky Strike's executives, and to whom he proposes marriage.

The next day, the main characters appear, embodying the cast. Salvatore Romano introduces a rough draft for a new Lucky Strike ad. The association of death with cigarettes is the conclusion of a report that Greta Guttman presents to Don, but the creative director throws it in the trash, manifesting his devaluation of the idea.

Don's next meeting involves Rachel Menken, the Jewish head of a department store. She has pretensions to upscale and establish her store as a Chanel fashion house, but Don suggests she offer coupons to housewives. Menken lashes out at the idea, and Don, angered, leaves the room.

The meeting with Lucky Strike executives, after a inefficient outset in persuasion, finishes with a incisive slogan constructed by Don exploring the idea 'Advertising is based on one thing: happiness': 'Everybody else's tobacco is poisonous, but Lucky Strike's is toasted.'

After the meeting, Roger Sterling urges Don to reconsider the opportunity to carry out public relations work for Richard Nixon, the Republican presidential candidate. A portrait is *painted* of Don as a man of success, who copes well with stress situations.

3) **Teaser**: End of the day and return to home: At the end of the pilot episode, Dan comes home after a busy day. Surprisingly—due to the flirtatious ambience shown—the spectators perceive that he is married and a father.

Story Analysis

Protagonist

Don is, professionally, characterized as an efficient, dedicated, and successful man. In spite of the lack of a solution the night before an important meeting, he, confronted with executives leaving the room, manages to produce a solution that pleased everyone.

Midge helps to qualify Don in the dialogue with him: 'Is this the part where I say, "Don Draper is the greatest ad-man ever and his big strong brain will find a way to lead the sheep to the slaughterhouse?"' (Weiner, 2006, p. 5).

Physically, he is handsome (Weiner, 2006, p. 1). In a psychological qualification, he is conservative (Weiner, 2006, p. 1). He is a disbeliever in love and, in spite of being married, is an adulterer.

Themes, Stereotypes, and Clichés

a) Dedication to work and competition: The depiction of workaholic men, dedicated to their professional lives and competing. Confirmation of the entrance of women in the professional world.

b) Gradual liberation of women in a period when women are gradually getting more power, independence, and freedom from of the social dogmas that consider them to have a secondary role compared to men. This theme is confirmed, for example, by: 1. Rachel, who has chief functions in her father's business, 2. Midge, an independent woman and Don's mistress. Don confirms the status of Midge: 'I'm serious. You have your own business and you don't care when I come over. What size Cadillac do you take?' (Weiner, 2006, p. 6); 3. Joan Harris is office manager and head of the secretarial pool at Sterling Cooper. On the other hand, Betty Francis embodies the stereotype of a woman dedicated to family.

c) Female objectification and conception of being a woman: Joan Harris embodies the femme fatal (stereotype). She uses clothes to define her silhouette and accompanies Peggy Olsen on her first day at the agency, advising how she should dress to capture male attention. Pete, observing Peggy's clothes, says to her: 'Are you Amish or something? (…) Well you're in the city, now. It wouldn't be a sin for us to see your legs. And if you pull your belt in a little bit, you might look like a woman' (Weiner, 2006, p. 16). The idea of woman being an object to admire is patent.

d) Sour conception of love: Insinuation that Pete will be married motivated by money and he personifies the womanizer (stereotype); once, on his bachelor night, he goes to Peggy's house, hoping to flirt.

Don says to Rachel:

'The reason you haven't felt it [love] is because it doesn't exist. What you call "love" was invented by guys like me to sell nylons. (…) I'm pretty sure about it. You're born alone, you die alone, and this world just drops a bunch of rules on top of you to make you forget those facts. But I never forget. (finishing drink) I'm living like there's no tomorrow, because there isn't one' (Weiner, 2006, pp. 47-48).

e) Don is married and has a mistress (stereotype).

Symbols

a) Cigarettes: symbol of a new era, in which smoking is synonymous with style. 'Smoking is the addiction more exclusively dependent on the vanity and charm' (Gikovate, 1987, p. 128). The cigarette symbolizes 'irreverence and independence, but in a measured, controlled way' (Gikovate, 1987, p. 128).

Final Considerations

The protagonists of the series are mostly men. The only exception is in *Homeland*, in which there is a co-protagonism and one of the principal characters (Carrie) is a woman. Additionally, only two are not married: Nucky (*Boardwalk Empire*) is widower and Carrie (*Homeland*) is single. They are also the protagonists who do not have children. Almost all of the central characters are successful in their lives, exhibiting good social positions and / or praise from hierarchies. The only exception is Walter (*Breaking Bad*), but the pilot episode shows a turnaround and a *new breath* in his life. Protagonists are humanized: endowed with feelings, confronted with doubts about life, and forced to deal with the challenges that life offers. They are not exceptional, nor are they perfect. Carrie, for example, goes outside the law to prove her suspicions. Brody (*Homeland*) lies. Don (*Mad Men*) is an adulterer. Nucky is seen to be cynical because his private and public behaviours contradict each other. Walter throws up an illicit adventure. Vital characters tend to be solitary, with the exception of the Lords: Lord Grantham (*Downton Abbey*) and Lord Ned (*Game of Thrones*).

None of the episodes makes use of a narrator, leaving to the film language the noble and powerful task of telling, advancing, showing, codifying, and constructing the story. Narrative, as Aristotle (2004) advocated, anchors in three principal structural moments: beginning, middle, and end. In the case studies analysed, the narrative is mostly chronological. The exceptions are *Boardwalk Empire* and *Breaking Bad*, whose beginning is already an advanced part of the story and is a key-event of the action, a twist. Generally, the beginning is an attractive passage / event, seeking to capture the spectator's attention to the story. In the two series mentioned, the use, early on, of a puzzling and frantic event, chronologically interrupted, captivates the viewer, who becomes curious and is stimulated to participate in a *time travel* to discover and accompany the antecedent sequence. In the case of *Game of Thrones*, the enigma and mystery, involving the strange creatures, function as a decoy. The middle, mostly, serves to introduce the rest of the elementary characters and their interactions, and the conflict. There is no rule that the protagonist appears at the start of the episode. In some cases (*Boardwalk Empire*, *Downton Abbey*, *Game of Thrones*), the spectator only meets the protagonist in the beginning of the middle, and suspense is maintained from the start about about the identity of the principal character. The end is a teaser to the next episode, launching and feeding curiosity and a wish to accompany the series.

The majority of scenarios is realistic, allowing better identification on the part of the viewer. The exception is *Game of Thrones*, which deals with a fantasized scenario. Also the time that story reports is mostly in the present or the twentieth century. The exception again is *Game of Thrones*.

The Manichaeism 'we / they' is present in all the pilot episodes analysed: *Boardwalk Empire*: legal / illegal; public / private acting; *Breaking Bad*: tedious / vibrant life; licit / illicit activities; *Downton Abbey*: nobility / servanthood; intra-group / extra-group; *Game of Thrones*: loyalty / betrayal; winners / losers; *Homeland*: heroism / anti-heroism; patriotism / betrayal; peaceful / terrorist; and *Mad Men*: partnership / competition; virtue / defect.

The theme of illegal activities is conveyed in *Boardwalk Empire* and *Breaking Bad*. The theme of succession of the estate in *Downton Abbey*. War, power, and succession in *Game of Thrones*. Terrorism, clash of cultures, and heroism in *Homeland*. And the 'mad men', success, and competition in *Mad Men*.

In terms of stereotypes associated with the protagonist, the gangster is explored in *Boardwalk Empire*, the law offender and the unpredictable in *Breaking Bad*, the good husband and father in *Downton Abbey*, the fearless and the warrior in *Game of Thrones*, the (anti)hero and heroine, and the (anti)patriot in *Homeland*, and the successful and the adulterer in *Mad Men*.

Table 5-2. Summary of the Pilot Episode's Narrative Structure Analysis

	Boardwalk Empire	**Breaking Bad**	**Downton Abbey**	**Game of Thrones**	**Homeland**	**Mad Men**
Protagonist (gender; marital status and parenthood, profession / occupation; stereotype; other attributes)	Male, widower, treasurer, the gangster, dual personality, powerful	Male, married and father, professor, worker at a car wash and then producer of meth, the law offender and the unpredictable, suffers a transformation from a weak and 'politically correct' man to a confident man, involved in illegalities	Male, married and father of three girls, Lord, the good husband and father worried about the family estate, powerful	Male, married and father, Lord, the fearless, the warrior and the observer of traditions, powerful	Male, married and father, retired US Marine Gunnery Sergeant, the hero *vs.* the villain / anti-hero and the anti-patriot (N. B.). Female, single, CIA intelligence officer assigned to the Counterterrorism Centre, the heroine, the patriot, and the Islamic convert, obstinate, breaks the rules to achieve her intents (C. M.)	Male, married and father, creative director of an ad agency, the successful and the adulterer, powerful, successful, adulterous

Theme	Illegal activities	Illegal activities, personal upheaval	Succession of the estate	War, power, and succession	Terrorism, homeland security, clash of cultures clash, and heroism	Success and competition
Manichaeism	Legal / illegal; public / private acting	Boring / vibrant lie; licit / illicit activities	Nobility / servanthood; intra-group/extra-group	Loyalty / betrayal; winners / losers	Heroism/anti-heroism; patriotism/betrayal; peaceful / terrorist	Partnership/competition; virtue / defect
Narrator	Absent	Absent	Absent	Absent	Absent	Absent
Division of Narrative Structure	3 parts: Climax / event-suspense; characters and interactions / 'rising action'; teaser	3 parts (1; 2.1, 2.2; 3): Climax; characters and interactions / 'rising action'; volte-face; teaser	3 parts: Presentation of geographical, temporal, and intimate environments; events-motor and characters and interactions /'rising action'; teaser	3 parts (1; 2.1, 2.2, 2.3, 2.4; 3): Event-suspense; characters and interactions / 'rising action'; teaser	3 parts: Presentation of one of the protagonists and information-motor; characters and interactions / 'rising action'; teaser	3 parts: Presentation of the protagonist; characters and interactions / 'rising action' during a day and a half; teaser

Narrative Time	Non-chronological or Non-linear–Flashback	Non-chronological or Non-linear–Flashback	Chronological or Linear	Chronological or Linear	Generally, chronological or Linear, with a time gap (ten months later) and Brody's flashbacks	Chronological or Linear (one night, one day and its night)
Type of Scenario	Realist (Atlantic City). Nucky is loosely based on the historical figure Enoch Lewis 'Nucky' Johnson (1883-1968)	Realist (Albuquerque, New Mexico, USA)	Geofiction (Downton Abbey)	Fantastic (fictional continents of Westeros and Essos, Winterfell, King's Landing–Capital of the Seven Kingdoms–, Pentos, Across the Narrow Sea)	Realist (Baghdad–Iraq; Washington, D.C.; Virginia—USA)	Reality (Madison Avenue, New York City, USA)
Time Period	1920's	Present day	Present day	End of a decade-long summer	Since 2011	1960's

Source: Own elaboration

References

Aristóteles (2004). *Poética*. Lisboa: Fundação Calouste Gulbenkian.

Aumont, J. & Marie, M. (2009). *A Análise do Filme*. Lisboa: Edições Texto & Grafia, Lda.

Barthes, R. (1966, 1977). Introduction to the structural analysis of narratives. In R. Barthes, *Image–Music–Text*. New York: Hill & Wang.

—. (1989). *Elementos de Semiologia*. Lisboa: Edições 70.

—. (2004). *Aula*. São Paulo: Editora Pensamento-Cultrix Ltda.

—. (2007). *Crítica e Verdade*. São Paulo: Perspectiva.

Bednarek, M. (2010). *The Language of Fictional Television: Drama and Identity*. New York: Continuum.

Botting, F. (2002). Culture, subjectivity and the real; or, psychoanalysis reading postmodernity. In B. Adam & S. Allan (Eds.), *Theorizing Culture: An Interdisciplinary Critique After Postmodernism* (pp. 87-99). London: Taylor & Francis.

Branigan, E. (1992). *Narrative Comprehension and Film*. N.Y.: Routledge.

Bulman, C. (2007). *Creative Writing: A Guide and Glossary to Fiction Writing*. Cambridge, UK; Malden, USA: Polity Press.

Darwin, C. (2010). *The Origin of Species*. Madison, Wisconsin: Cricket House Books.

Daniel, D. (2013). 7 Management Lessons from Downton Abbey. *mendix*. Retrieved from: http://www.mendix.com/think-tank/7-management-lessons-from-downton-abbey/

Feuer, J. (1986). Narrative Form in American Network Television. In C. MacCabe (Ed.). *High Theory/Low Culture: Analyzing Popular Television and Film*. Manchester: Manchester University Press.

Gansa, A., Gordon, H. & Raff, G. (2010, August 16) (teleplay by). *Homeland. Writers' draft*. Developed for American Television by Howard Gordon & Alex Gansa. Based on the Keshet Broadcasting Series "Prisoners of War" created by Gideon Raff. Retrieved from http://1e36a764da2f6b46c156-0fa6f106d65 4e15326ee14d2a07c02d0.r64.cf1.rackcdn.com/2013/07/Homeland_1x01_-_Pilot.pdf

Gikovate, F. (1987). *Vício dos vícios: um estudo sobre a vaidade humana*. São Paulo: MG Editores Associados.

Guiraud P. (1999). *A Semiologia*. Lisboa: Presença.

Herman, L. & Vervaeck, B. (2005). *Handbook of narrative analysis*. Antwerpen: University of Nebraska Press.

Hooyman, N. R. & Kiyak, H. A. (2008). *Social Gerontology: A Multidisciplinary Perspective*. Boston: Pearson Education.

IMDb (2008). Quotes for Jesse Pinkman (Character) from 'Ruptura Total'. *IMDb*. Retrieved from http://www.imdb.com/character/ch0096617/quotes

—. (2011). A Guerra dos Tronos (TV Series) Winter is Coming—Quotes. *IMDb*. Retrieved from http://www.imdb.com/title/tt1480055/quotes

Jacobsen, C. (2006, October 6). How TV Met Narrative Sophistication. *flowtv.org*. Retrieved from http://flowtv.org/2006/10/reunion-the-nine-24-narrative-flashback-arrested-development-the-office-how-i-met-your-mother/

Joly, M. (2003). *A Imagem e a Sua Interpretação*. Lisboa: Edições 70.

Julien, N. (2012). *The Mammoth Book of Lost Symbols—A Guide to the Language of Symbolism*. London: Constable & Robinson Ltd.

Kubrick, S. (1968) *2001: Space Odyssey*. United States of America; United Kingdom: Metrocolor, Metro-Goldwyn-Mayer, Stanley Kubrick Productions, 142 mins.

Levin, G. (2011, September 28). Showtime's 'Homeland' skeptical of war on terror. *usatoday.com*. Retrieved from http://usatoday30.usatoday.com/life/television/news/story/2011-09-29/showtime-homeland/50594388/1

Lexicon, H. (1990). *Dicionário de Símbolos*. São Paulo: Editora Pensamento-Cultrix Ltda.

Maquiavel, N. (2007). *O Príncipe*. Lisboa: Guimarães Editores.

Mendes, I. A. (2006). *Números: O Simbólico e o racional da história*. São Paulo: Editora Livraria da Física.

Metz, C. (1974). *Film Language: A Semiotics of the Cinema*. Chicago: University of Chicago Press.

—. (1982). *The Imaginary Signifier: Psychoanalysis and the Cinema*. Bloomington: Indiana University Press.

—. (1988). *Linguagem e Cinema*. São Paulo: Perspectiva.

Penafria, M. (2009). Análise de Filmes—conceitos e metodologia(s). *BOCC*. Retrieved from http://www.bocc.ubi.pt/pag/bocc-penafria-analise.pdf

Perrine, L. & Arp, T. R. (1983). *Literature: Structure, Sound, and Sense*. New York: Harcourt Brace Jovanovich.

Pointz, S. *Visual Storytelling and Narrative Structure*. Burnaby, B.C.: Open Learning Agency, Pacific Cinémathèque. Retrieved from http://education.cinematheque.bc.ca/pdfs/f_h_guide03.pdf

Propp, V. (1965; 1970). *Morphologie du Conte*. Paris: Le Seuil.

Richards, K. G. (1992). *The Gettysburg Address*. Chicago: Childrens Press.

Segal, D. (2011, July 6). The Dark Art of 'Breaking Bad'. *The New York Times*. Retrieved from http://www.nytimes.com/2011/07/10/magazine/the-dark-art-of-breaking-bad.html?pagewanted=all&_r=0

Seitz, M. Z. (2010). 'Boardwalk Empire': Episode One. *New Republic*. Retrieved from http://www.newrepublic.com/article/77765/boardwalk-empire-episode-one#

Shapiro, K. J. & Beum, R. L. (1965). *A Prosody Handbook*. New York: Harper & Row.

Stott, J. (2013, April 10). Inside the mind of Don Draper—understanding his marketing philosophy. *guardian.co.uk*. Retrieved from http://www.guardian.co.uk/media-network/media-network-blog/2013/apr/10/marketing-don-draper-mad-men

Swenson, J. E. (2000). Action Plan for the Conservation of the Brown Bear in Europe (Ursus Arctos). *Nature and environment*, No. 114. Council of Europe: Council of Europe Publishing.

VanDerWerff, T. (2010, September 20). "'Boardwalk Empire' recap: Meet Nucky

Thompson". *latimes.com*. Retrieved from http://latimesblogs.latimes.com/showtracker/2010/09/boardwalk-empire-recap-meet-nucky-thompson.html

Vanoye, F. & Goliot-Lété, A. (2006). *Ensaio sobre a análise fílmica*. Campinas: Papirus.

Walter, M T. M. T. & Baptista, S. G. (2007). A força dos estereótipos na construção da imagem professional dos bibliotecários. *Informação & Sociedade: estudos*, João Pessoa, 17(3), 27-38, set./dez.

Weiner, M. (2006). Mad Men. One-Hour Pilot "Smoke Gets in Your Eyes". United Kingdom: *raindance.co.uk*. Retrieved from http://www.raindance.co.uk/site/picture/upload/mad-men-1x01-smoke-gets-in-your-eyes.pdf

Werness, H. B. (2006). *The Continuum Encyclopedia of Animal Symbolism in World Art*. New York: Continuum.

Williams, M. (1986). *Patterns of Poetry: An Encyclopedia of Forms*. Baton Rouge: Louisiana State University Press.

Zillman, L. J. (1966). *The art and craft of poetry: an introduction*. New York: Macmillan.

CHAPTER SIX

POWER, MONEY AND CRIMINALITY IN THE NEW BULGARIAN TV SERIES

VALENTINA GUEORGUIEVA

Since 2010, the Bulgarian media sphere has witnessed the birth of a wave of TV series which have started a new tradition in the production of film and media content. After *Home of Glass*, broadcasted during 2010–2012 by the commercial TV channel bTV, a couple of production companies launched the filming of a number of TV series, commissioned for the most part by commercial channels. What distinguishes these productions from previous attempts at creating popular TV series in Bulgaria is their plot and production history. Therefore, I will consider two main traits as distinctive of what I call here the new Bulgarian TV series: the first is related to the plot and narrative structure; the second to the economic and creative actors responsible for the production of the series.

Without exception, the storyline of all the new Bulgarian TV series is staged in the social and political context of contemporary Bulgaria. With frequent references to actual events, these series relate a fictional story (be it a love story, a family saga, a criminal investigation) navigating in the realm of a pseudo-reality constructed by the same events covered in the news programme of the TV channels broadcasting the series in question. As far as the second main trait is concerned, unlike the TV series of the previous periods, produced by the major Bulgarian film studio (Boyana film studios) in cooperation with the national television (BNT), the new TV series are the result of the collaborative efforts of established advertising companies (with their studios, equipment, and staff) and newly emerging scriptwriters. The producers and directors of the series were not previously established as film directors, nor has their career begun in the film industry. Their names have emerged from the fields of advertising and production of TV shows. In what follows I will consider these two main features of the new Bulgarian TV series.

Starting with the second trait, who are the producers of the new

Bulgarian TV series? A number of names appear repeatedly: Dimitar Gochev, Dimitar Mitovski, Magarditch Halvadjian, and two or three more names, all connected with the companies SIA advertising, Global Films, Dream Team Productions, and Old School Productions.

The three most popular new TV series—the groundbreaking *Home of Glass* (in Bulgarian *Staklen Dom*, aired by bTV, 2010-2012); the most praised by the critics, *Under Cover* (*Pod Prikritie*, commissioned and broadcasted by the national television, 2011-2013); and the successor of *Home of Glass* oriented towards a younger audience, *Revolution Z: Sex, Lies and Music* (*Revolutsia Z*, bTV, 2012-13)—, are all the product of Camera Ltd. and SIA Advertising. These joint production companies are also in charge of the most recent TV saga *The Family* (bTV, starting form April 2013). The founders and owners of the companies in question are the creative team Dimitar Gotchev and Dimitar Mitovski, who also have the merit of being the producers of these four series.

Dimitar Gotchev started working as an electrician and later cameraman in Bulgarian National Television in the 1980s. He later acquired a degree from the National Academy of Theatre and Film Arts (NAFTA) and was engaged in Boyana Film Studios. In the first years after the democratic changes he founded SIA Advertising, one of the leading companies producing television commercials, and the closely connected production company Camera Ltd. Nowadays, he is most renowned as a producer of TV series, but he also works on documentaries (*Stoichkov* and *The last Ambulance in Sofia*) and feature films (*Ave*, released in 2011, and *Hunting Down Small Predators*, 2010)[1]. The most viewed of his works are, however, the numerous advertisement clips that he has directed, listed in his profile on the web-site of Camera Ltd. He is one of the former television professionals who turned to the more profitable business of television advertising during the 1990s. Later, after acquiring capital and status, he could turn to the production of films and TV series, starting from 2010.

His partner and co-founder of SIA Advertisement and Camera Ltd., is Dimitar Mitovski, who studied to be a film and animation director at NAFTA in the 1990s. Apart from his work as a producer for the TV series listed above, with Dimitar Gochev, he became known as the director of the feature movie *Mission London* (2010), acclaimed as the first Bulgarian

[1] The data about the academic background, career and productions of Dimitar Gochev are retrieved from his pages on the web-site of "Camera Ltd.", Wikipedia and IMDb, see http://www.camera.bg/directors/39, http://bg.wikipedia.org/wiki/Димитър_Гочев, and http://www.imdb.com/name/nm1189479/

blockbuster. Based on the novel of Alek Popov telling the story of hoaxes and abuse at the Bulgarian embassy in London, the film is a light comedy aspiring to be an 'eccentric political satire',[2] and feigning the scale of a Hollywood movie. Mitovski's profile on the web-site of Camera Ltd. promotes him as 'one of the most prominent figures in Bulgaria's advertising market.'[3] After the success of *Home of Glass*, bTV commissioned another series—*Seven Hours of Jet Lag* (*Sedem chassa razlika*, 2011-2013)—from a different production company. Global Films is owned and directed by Magarditch Halvadjian. According to the short biographical notice on Wikipedia, he studied circus arts in Moscow, and then film directing in a private school. He is known to the public from the TV shows that he produces and appears on (*Got Talent*, *Lords of the Air*—the Bulgarian version of *Striscia La Notizia*—, *Complete madness*, *Russian roulette*). The series *Seven Hours of Jet Lag* is not his first attempt in film production, though definitely the most well known of his works and one of the most avidly followed series by the audience. Among the productions of Global films are popular TV shows, a couple of completely unknown titles of feature films (mainly from the 1990s), music videos, and a huge number of commercials.[4]

The list of popular series continues with, in order of appearance, *Sofianites More Than Enough* (*Stolichani v poveche,* bTV, 2011, Dream Team Productions), *Staged Property* (*Etazhna sobstvenost,* NOVA TV, 2011, Hidalgo Productions), *Where is Magi?* (*Kude e Magi?*, bTV, 2012-13, bTV studios), *The Repayment* (*Otplata*, NOVA TV, 2013, OldSchool Productions), and *The Tree of Life* (*Durvoto na zhivota*, TV7, 2013, Dream Team Productions). The production companies Dream Team Productions, Hidalgo, and Old School Productions are definitely not new names in the field. Though they have only two or three TV series in their lists of assets, they are the leading producers of TV shows and games (Old School), reality formats and TV comedies (Dream Team).[5]

To cut a long story short, it may be affirmed that the new Bulgarian TV series are the offspring of the advertising industry. Especially in the first episodes of the first series—*Home of Glass*—one can easily see the connection with TV commercials in the techniques of filming, the camera

---

[2]According to the description of the movie in Mitovski's profile on IMDb: see http://www.imdb.com/name/nm1190360/bio

[3] http://www.camera.bg/directors/32; for Mitovski see also http://bg.wikipedia.org/wiki/Димитър_Митовски

[4] Data retrieved form http://www.globalfilms-bg.com/?go=ads&lang=en.

[5] For more detailed information see http://www.oldschool.bg/# and http://www.imdb.com/company/co0305713/

work and the lighting, the set decoration, scenography and costume design. Besides, from the length and continuity of scenes, from cuts and film editing, the spectator got the impression that he was viewing a prolonged version of a TV commercial. With the advance of the plot and the number of episodes aired, but no doubt also with the lengthier experience of filming and hence the growing skills of the filmmakers, this impression was lost. Nevertheless the subjects and characters in frame remain the same—beautiful happy people, nice clothes, sophisticated home and office design, fancy cars, and glossy images. This is exactly the same perfect world that we know from advertisement clips. The pictures in the new Bulgarian TV series still look like promotional videos, selling you a perfect world. But what do they promote?

This last question points to the second trait of the new Bulgarian TV series, related to their plot and narrative structure. They use an imagery of luxury and comfort to mirror characters and events from the political and economic news, and also from the criminal chronicles of contemporary Bulgaria.

The same figures, scandals, backstage political games, corporate alliances, and criminal deeds are narrated in the film plot. The spectator can easily recognize references to real-life events, such as the business with kidnapping whose peak of activity was in 2009, the street protests in defence of the natural reserve Strandja in the summer of 2007, the presidential campaign of a former mayor of Sofia, and other recent events.

The storylines of the series are different, but they are all set in the neat interiors of bourgeois homes, of luxury restaurants and offices, expensive cars, bars, and shops. *Home of Glass* is like an advertisement for the mall where it was filmed. It pictures the lifestyle of the owners of a shopping mall, their families and teenage children. The characters in the series spend all their lives inside the mall—they work there, spend their free time there, quarrel and make up there. This is their home. The title itself is a reference to the shiny glass walls of a contemporary temple of consumption, but also a metaphor for the fragility of human relationships and of the instability of the family institution in a world of individualization and consumerism.

Under Cover and *Seven Hours of Jet Lag* display the bars, homes, offices, and expensive cars of the emblematic figures of organized crime. They share the same style and way of life with businessmen, political elites, and high officials of the judicial system. The relationships between organized crime, judges and politicians are not hidden, but exhibited to the point that they are no longer surprising to the public. They are all members of the class of rich and influential people. Their lifestyle is the thing that unites them, according to the series.

Actually all the series picture the lives of the rich, be it the owners of a mall (*Home of Glass*), well-to-do female judges, journalists, and their gangster friends and lovers (*Seven Hours of Jet Lag*), owners of a big holding in the construction business or of an architectural office (*Where is Magi?*), gangsters, police officers, and politicians (*Under Cover*). Even *Sofianites More Than Enough*, which claims to be the more folksy version of the above, telling the story of two families living the ordinary lives of country people, is about the family of the mayor of a small town near Sofia and their rival clan—the owners of the town's hotel. Only the series *Staged Property* is about the humble inhabitants of an apartment building, who belong to the popular classes. But by season three the action moves to the environment of a gated community with a pool. In a word, the new Bulgarian TV series stand in line with the logic and expressive style of TV commercials, whose function is to create desires by presenting to the audience the fantastic life of wealthy people.

Who are the characters inhabiting this world of dreams? People with power—the big corporate sharks, often connected with dirty deals, the very crime bosses themselves, politicians and businessmen. The series do not attempt to portray these characters in a better light. On the contrary, the links between the business world and the world of organized crime are in full view. The weakness and dependence of politicians under the pressure of corporate interests and gangster threats are narrated and explained according to the logic of 'money creates influence' and 'with more money you can buy anything.'

In *Home of Glass*, the main shareholder, Dimitar Kassabov, deceives his partners and founds a new competitor mall. The manager, Nikolay Zhekov, is caught in a financial fraud and throughout season two is in jail. When he is eventually released he sinks again into half-criminal 'business initiatives' and dirty deals. The only character presented as honest, noble, and kind-hearted is the elder son of the founder, Kamen Kassabov, who comes back to Bulgaria after years of emigration in the USA. The scenario builds on his moral integrity until it turns out that in the past, before he emigrated, he was implicated in criminal schemes of trafficking women. The whole story of 'the respectful businessmen' and their family saga is set against a background of criminality, exemplified by the unnoticed but constant presence and tangible threat from the mafia boss Methody Ganev.

In the other most watched series on bTV, *Seven Hours of Jet Lag*, the connections between business and criminality, but also with the government and the judicial system are the building blocks of the storyline. The intrigue is centred around some areas of land at the seaside.

These are included within the Strandja nature reserve and therefore protected by the state, but they form an object of interest for two conflicting groups of 'investors'. One of the sides in this conflict is taken by the intimately related 'corporate interests' of a criminal boss (Miloto) and a corrupt judge (Tanya Stoeva who is in a relationship with Miloto). In this couple, the judge ensures the 'administrative compatibility' of the mobster's plans for constructing a complex of hotels in the nature reserve. On the other side of the conflict, the plot features the group around the former state security general Luka Karlov (the name replicates the former socialist politician Karlo Lukanov, connected with the structures of the state security who acted in favour of Russian interests; he was shot by the mafia at the beginning of the 1990s). The group around the general wish to get their hands on the same land in order to ensure a future pipeline route. In between these two conflicting groups, we see an investigative journalist (Svetlana Stefanakova) who is moved by unaffected professional desire to reveal the fraud, but falls into the pitfall of her own dark roots and lust—she is the daughter of the general who pulls the strings, and is involved in a long term affair with one of his lieutenants (Cyril Kotzev).

In 'Where is Magi?' the plot is based on the disappearance (or supposed kidnapping) of the teenage daughter of the owner and manager of a construction holding, Radoslav Tabakov. During the investigation it turns out that for the past years he has collected a solid portfolio of enemies. In an attempt to rehabilitate his honesty, his sister and second most important figure in the holding, Katerina Andonova, affirms that it is impossible to reach the top and still have 'clean hands'. Although in this series there is no explicit linking of business with crime and politics, it puts forward the same message—that there is no businessman with an unblemished reputation.

If there are any honest people in these series, they inevitably suffer from inner doubts and struggles with their own vicious intentions. The decent characters are put to moral tests, such as infidelity or obsessive love for a morally corrupt person. In *Where is Magi?*, the virtuous wife, Vyara Tabakova, has an affair with the investigating officer, Andrey. In *Seven Hours of Jet Lag*, the unbribable public prosecutor Ivanov and the devoted assistant Donchev are tragically and sacrificially in love with the corrupted judge Tanya Stoeva.

Another, rather comic, storyline is developed with the good policeman in *Undercover* and the respectful doctor in *Where is Magi?*—their 'punishment' is their rebellious teenage children. Dressed in a subcultural code, these youngsters are strongly inclined to the easy side of life—unrestrained pleasures, light drugs, easy fame. It does not become clear from the series what their rebellion is about, if it is not just to annoy their parents.

In general, the images of youth in the new Bulgarian TV series are presented in the key of lighthearted pleasures and easy life. The usual worries of their 'regular' life are reduced to shopping in the mall, evenings and nights spent in nice bars and discotheques, a breathtaking ascent to media fame (for Alex *in Home of Glass*), and aspirations for the rock scene (for the heroes *in Revolution Z*). Their prospects for the future are unproblematic: they have the financial stability of their parents and the promise of an easy life of consumption and pleasure. It seems that there is nothing better a young person could dream of. The series do not offer to the younger audience alternative models of conduct.

If there could be one single conceptual frame for the female characters in the new Bulgarian TV series, this could be the image of weakness and fragility. In the series *Under Cover*, dedicated to the man's world of criminality and the police, the female characters are only sketched. Sunny, the oppressed and humiliated young lover of the mobster, who tries to escape from him and collaborates with the police, is killed before the end of the first season. The anxious and sick old mother of the boss's right hand man (and later rival) only appears a couple of times throughout the series and has no influence on the development of the story. The boss's underrated and misjudged female advocate is humiliated, treated as a sexual object, and eventually killed by him. The message seems to be that in this harsh world of men there is no place for such fragile creatures. In the world of money and corruption the only women who can compete with men are hardened and devoid of moral sense like the judge Stoeva, or like the business bitches Katerina Andonova (*Where is Magi?*) and Elena Atanasova (*Home of Glass*), who also have their brothers or husbands in the same business to protect them. Forced to join in the game after the (pretended) death of her husband Dimitar Kassabov, Boryana, in *Home of Glass*, is constantly on the verge of an emotional collapse, until she is rescued by her new husband Kamen Kassabov. In *Seven Hours of Jet Lag*, the image of the investigative journalist Svetlana Stefanakova is also developed in the key of typical female weaknesses—she is unstable, anxious, emotional and volatile, inconsistent, and unfaithful.

It is hard to find a female character who does not enter into the scheme of patriarchal stereotypes about femininity, and who can offer a role model of the independent woman. To some extent this is Siana in *Home of Glass*. She is a single mother and financially independent due to her own educational background and professional skills. But she is a victim of her own 'devils'—she cannot choose a man and constantly hesitates between the biological father of her child and the man who loves and cares for her. She is unstable and creates a mess as a result of her hesitations in love. Eventually, it is the two men who decide for her who should be the man by her side.

It is only in *Home of Glass* that the audience can find representations of the ethnic minorities traditionally living in Bulgaria. Their inclusion in the storyline is a response to the real-life hostility between the Roma and Bulgarian population, whose occasional outbreaks are always a news story. In the series, Dany Kassabov, the younger son of the family of mall-owners, becomes a victim of his own ethnic prejudices, and rejects the candidature of a Roma girl with an excellent academic background applying for a position in his office. The girl's brother is moved by her disappointment and decides to give Dany a lesson, a gentle one. But Dany does not stop there, the conflict escalates and the Roma boy is badly injured and spends some time in coma. He is healed after the intervention of the generous and considerate Boryana, Dany's mother. Dany confesses and receives absolution for his sin from the Roma family, touched by the kindness of his mother. It seems, however, that the scriptwriters are uneasy about the representations of ethnic conflicts and social distance in the film. The conflict is resolved somehow in accordance with the directions of Christianity—generosity, mercy and forgiveness are the turning points. The moral is too obvious: 'You can only be a victim of your own prejudices.' This episode hardly contributes to a better understanding of the problems of social distance and discrimination against Roma. Neither does it create a better image of the minority.

The presence of people of different sexuality is also marginal in the new Bulgarian TV series. We can only see such examples in *Seven Hours of Jet Lag*. In the first season, this is the MP who is seduced on the orders of Tanya Stoeva by a fitness instructor. Though he never enters their secrete scheming with Miloto and remains a person with a pronounced moral stance against criminality, he is humiliated and physically aggressed by the mobster. He is never rehabilitated in the eyes of the audience later in the development of the storyline. The script likewise treats the relationship of Nina Rossi and Olesya (in the American part of the same series) in a negative key. Nina succumbs to the perverse fascination of Olesya, and the girls conspire together to deceive both the Russian mafia boss Rasputin and the Bulgarian mobster Miloto, and to escape with untold wealth. In both cases, homosexuality is used by the scriptwriters to create a negative impression of the characters. Instead of creating a positive image of people of different sexuality, the series perpetuates negative stereotypes about homosexuality as something vicious and reprehensible.

To sum up, what is the overall message of the new Bulgarian TV series? We are left with the impression that they are nothing but the mirror of reality, that all they do is to use some fictional characters to relate the actual story of contemporary Bulgaria—the same political matters, the same business interests in malls and the construction business, the same kidnaps and bribery, even the same stereotypes and prejudices. The familiar stories from the news are inserted into the fantastic houses and office interiors that we know from the commercials. But this simple trick of the advertisement industry is slippery—it presents 'the real' as desirable. Actuality is transformed into a phantasm.

The new Bulgarian TV series have the function of a typical promotional video—they create desires. They tell us what to dream about, what to long for. No effort is made to create role models that go beyond the ideal of consumption. Alternative fantasies are not on offer. Different young people, minority groups, and people of different sexuality are either absent or presented in their stereotyped image. It seems that there is no way out of the world of consumption and existing inequalities.

What is more striking is the permanent assertion in the series that 'there is no fair business,' that politics is a 'dirtygame', that politicians are subjected to the interests of business, and that business is subjected to organized crime. The inherent suggestion is that criminality, corporations, and politics have become one and the same and that there is no way out of this situation. There is no alternative to this malicious unity.

The reaffirmation of the trinity of power, money, and criminality on the level of the phantasmic imagery of the TV series is dangerous for two reasons. Firstly, because it reinstates the illegal and immoral connections between rich and influential people as something inescapable, inevitable, and normal. It thus gives it the status of normality and blocks the political imagination that could find possible alternatives and work for social change. Secondly, it presents the lifestyles of rich (and criminal) people as something desirable—they are the beautiful happy people of the commercials. The reality of a malicious trinity slips into the realm of the fantastic. Lured by the bliss of their lifestyle, the audience forgets about the sources of this fantastic wealth and is caught in the trap of an impossible dream.

References

Creeber, G. (Ed.) (2001). *The Television Genre Book.* British Film Institute/BFI Publishing.

Foucault, M. (1975). *Surveiller et punir. Naissance de la prison.* Gallimard: Paris.

Johnson, M. L. (Ed.) (2007). *Third Wave Feminism and Television: Jane Puts It in a Box.* I.B.Tauris & Co: London.

McRobbie, A. (2009). *The Aftermath of Feminism: Gender, Culture, and Social Life*. Sage Publications: London.

Mittel, J. (2004). *Genre and Television: From Cop Shows to Cartoons in American Culture.* Routledge: London.

Newcomb, H. (Ed.) (2000). *Television: The Critical View* (6th edition). Oxford University Press: U K.

CHAPTER SEVEN

RECONSTRUCTING HEALTH: PERCEPTIONS AND REPRESENTATIONS OF MEDICAL TV SERIES

VALENTINA MARINESCU

Introduction

One can notice that the institution of medicine has long been a favourite subject for TV series. Commenting on the impetus for the ever-rising popularity of medical dramas, Zurawik (1999) pointed to modern culture's obsession with health consciousness combined with widespread anxiety about health care delivery in health maintenance organizations and other forms of managed care (Zurawik, 1999).

Television medical dramas have enjoyed popularity for several decades. Studies on these popular programmes have investigated how accurately they reflect actual medical practices and situations, as well as their effect on attitudes, knowledge, and behaviours (Czarny, Faden, Nolan, Bodensiek, Sugarman, 2008; Quick, 2009; Wicclair, 2008). Communication researchers have studied whether health-related information in medical dramas improves health-related knowledge, attitudes, or behaviours (Brodie et al., 2001; Hether, Huang, Beck, Murphy, Valente, 2008; Morgan, Movius, Cody, 2009).

However, researchers have paid little attention to how and why people watch medical dramas and what factors influence those motives and consequent behaviours.

The present article investigates the patterns of watching TV medical dramas in Romania among doctors and students of the Faculty of Medicine in Bucharest.

'What are the main patters of watching medical TV dramas among doctors and students of the Faculty of Medicine?'—this is the main research question to which I seek an answer.

Theoretical background

Social representations theory

Social representations theory (Moscovici, 1984; Moscovici, 1988; Moscovici, Hewstone, 1983) implies that there is a progression from mass media representation to cognition to interpersonal communication to social representation, but other researchers assert that the progression is by no means unidirectional (Joffe, 2003; Lupton, 1994).

Social representations theory has been used widely as a loose theoretical framework over the past few decades. Although social representations theory does not have the straightforward predictive power of many of the cognition-based theories currently favoured by social scientists, such as the theory of reasoned action (Fishbein, Ajzen, 1975) and the elaboration likelihood model (Petty, Cacioppo, 1986), its breadth may ultimately enrich and even surpass the explanatory power of other theories. Social representations theory at its most basic states that social and individual views of a topic will be a function of a combination of mass media framing, individuals' cognitions about the topic (largely a function of information provided by the mass media, though heavily influenced by the commonsense of an individual's community regarding similar topics), and interpersonal, everyday communication regarding the topic.

The influence of an individual's own communities on patterns of the development of social representations should be taken into account. Social representations do not necessarily develop in some monolithic form throughout society as a whole. Thus, the social distribution of knowledge depends on the social groups to which people belong (Flick, 1998). In the case of worksite organ-donation campaigns, information disseminated to the social network represented by an organization's employees will likely remain largely within that network. In fact, social groups themselves can be defined by the content of their 'everyday knowledge'. However, knowing how people acquire information is as important as knowledge about their specific beliefs (Flick, 1998).

Kruglanski (2001) recommends that studies of social representations focus both on a wide variety of topics and on how a single topic is represented within diverse communities. Differences in knowledge (or memory of particular facts) will vary among groups because of differences in norms, values, and identities (Sommer, 1998).

Cultivation Theory

According to cultivation theorists, television is a storyteller (Morgan, Signorielli, 1990, p. 13); television is the main proprietor of cultural stories and messages. Although many communication theories and studies focus on the short term behavioural effects of exposure to media content (Perse, 2001), cultivation theory proposes a much more pervasive effect: enculturation (Gerbner et al., 2002).

The main idea of cultivation theory is that television cultivates shared conceptions of reality among different social groups (Gerbner et al., 2002).

Two of the main processes of cultivation theory are resonance and mainstreaming. Resonance refers to an amplification of cultivation for a certain group, because television's depictions about a topic are congruent with that group‘s everyday realities (Shanahan, Morgan, 1999). According to Shanahan and Morgan (1999), resonance arose from criticism of some empirical findings.

Uses and Gratification Theory

A focus on audience motives and activity is the hallmark of the uses and gratifications perspective (Katz, Blumler, Gurevitch, 1974). Empirical studies have demonstrated strong associations between gratification sought and obtained from media use (Palmgreen, Wenner, Rayburn, 1980), and between audience activity and gratification sought and/or obtained (Levy, Windahl, 1984). If this holds for medical dramas, gratifications by viewers sought might also shape audience activity and gratification obtained from medical drama viewing, including whether health-related information is learned.

Research has explored how media narratives lead to viewers' health-related attitudes or behaviours (Hinyard, Kreuter, 2007; Slater, 2002), and medical dramas often present health-related information (Hether, et al., 2008; Morgan et al., 2009). Examining how viewers' motives for watching medical dramas influence their behaviours is the next logical step, necessary for understanding how to deliver health information via medical dramas more effectively.

Uses and gratifications is a perspective employed to examine how audiences use media, positing audience members as active: each audience member selects media in such a way as to fulfil needs or wants (Rubin, 2009). Past research has investigated viewers' motives toward entertainment programs such as soap operas (Babrow, 1987; Rubin, Perse, 1987) and reality shows (Nabi, Biely, Morgan, Stitt, 2003; Papacharissi,

Mendelson, 2007). Because entertainment media are basically for enjoyment, motives for diverse genres tend to be similar: entertainment, relaxation, social interaction, and habit. These likely also apply to medical dramas.

However, each type of entertainment media also has distinct motives associated with it. One possible motive for viewing medical dramas is health information seeking. Information seeking has been studied as a motive for television news viewing (e.g., Perse, 1990; Vincent, Basil, 1997), but few studies have been conducted on information seeking as a motive for viewing entertainment television. Although learning has been discussed as one of the motives for soap opera viewing, research has only explored learning about social and emotional situations (Babrow, 1987; Livingstone, 1988; Rubin, Perse, 1987), not about information on a specific topic.

Published research on acquiring health information has shown that people can gain health information by passive exposure to health information through routine media use or conversation (i.e. information scanning), as well as by actively engaging in health information seeking (Hornik, Niederdeppe, 2008; Shim, Kelly, Hornik, 2006). Information scanning is considered to occur without strong motivation or interest in obtaining information; however, it influences knowledge (Hornik, Niederdeppe, 2008; Niederdeppe et al., 2007; Zukin, Snyder, 1984). When information scanning is applied to medical drama viewing, medical drama viewers' routine media use would be expected to lead people to learn health-related or medical-related information. Empirically, health information learning has been demonstrated as a result of medical drama viewing (Davin, 2003).

However, whether the health information learning from medical dramas can be a gratification sought in viewing the dramas has not been studied.

Methodology

As the research objectives I have the following questions:

Q1. What are the audience's motives for watching medical dramas?

Q2. Is medical drama viewing associated with health information orientation for medical students and staff?

To answer the research questions I conducted a set of seventy-two interviews with family doctors and students of the Faculty of Medicine in Bucharest. The data collection took place in February-May 2013 in Bucharest.

Results Analysis

When asked what the medical TV series they watch, 52 of the respondents indicated *Grey's Anatomy*, 44 respondents referred to *House, M.D.*, and the TV series *Scrubs* was nominated only by twenty respondents from the total sample. The high audiences of *House* and *Grey's Anatomy* are not a surprise at the level of the general public (Barnet, 2007) and the results of other researches were confirmed (Quick, 2009). But what were the motives for watching those medical TV series in the case of Romanian doctors and students of the Faculty of Medicine?

The main reason to watch these series is to obtain more information about the medical domain: illness, treatments, diagnostics, and drugs. The second reason was not entertainment but to get a deeper understanding of the functioning of health systems in other countries, in order to facilitate a comparison with the Romanian state of affairs in medical domains. The third reason invoked by the respondents was to facilitate communication with peers—that is, other family doctors and/or other students of the Faculty of Medicine.

S5: I watch this TV series to see what the [medical] situation in other countries is like, if it is similar with what is now in Romania.

S22: Grey's Anatomy is a TV series which shows what the feelings of a young doctor are. I found it interesting to see that and compare it with my own situation.

S30: Because each episode brings to us a lot of information regarding pathology, it helps us in keeping up-to date with information in the medical domain.

S41: In general I find this TV series [Dr. House] interesting, full of situations one can meet in everyday practice. At the same time, it shows us the way in which doctors and patients discuss and how the doctors discuss one to another. And some of these situations can also be found in Romania.

Thus the TV series fulfil a general informative function, a social-cohesion one, and an educational one, thereby validating the theory of 'uses and gratification' associated with media message consumption (McQuail, 1993, p. 73).

'Professionalism', 'team-values', and 'human-values' are the most frequent elements of the TV series' script that are most appreciated by Romanian doctors and students of the Faculty of Medicine.

S14: In "Grey's Anatomy" I like the team spirit which aims to better serve the patient. The whole team works so that the patients can have the best medical treatment. One can also observe lot of dynamism. The relations at the work-place are very well described in the TV series

S22: The values which are at the core of the TV series are basic human values.

S69: The professionalism is something one cannot find in Romania. I also like the humour of the main character, and the hospitals, which are above Romania as regards their equipment and their clinical procedures.

The combination between the fictional character of the specific cases presented in medical TV series and the high level at which the medical system is presented in those drama is the main characteristic emphasized by the respondents in answer to the question: 'To what degree do you consider that the medical TV series resemble everyday reality in the Romanian health system?'

S6: I suppose that they are very similar with what is now in the US medical system...unfortunately in our country doctors haven't so much funding and they cannot approach medical science in the way presented in these TV series.

S28: It is difficult for me to judge. I think that maybe they are like the situation in the USA, where the TV series were produced. But I also think that there is a lot of fiction in these TV series. Not all medical cases are exceptional in daily practice, because in general a doctor treats a common illness.

S71: I think that they are somehow similar to real life. I appreciate the fact that I could find in those TV series some treatments and some drugs known from lectures at the Faculty. For example, in one episode a patient had liver-problems and Dr. House recommended Silimarin to him. This is exactly what a person with this kind of illness should take as a drug.

Conclusions

As pointed out in a number of studies (Freimuth, Stein, Kean, 1989; Johnson, Meischke, 1993; Nelkin, 1995), these health-related messages are effective devices of 'social learning', the people being able to learn what it means to be healthy with the help of media (Freimuth, Stein, Kean, 1989; Johnson, Meischke, 1993; Nelkin, 1995). Moreover, public health policies can be modified sometimes by the transmission of messages relating to health (Cho, 2006). Incorrect or unrealistic information from medical reports may mislead the people working in public health policy and may contribute to the admission of unfair laws and regulations (Hotz, 2002; Wells, Marshall, Crawley, Dickersin, 2001), the latter being regarded by many scientists as true threats to public health (Voss, 2002). Health-related news from the media has functioned largely as a factor that can influence the actions of doctors, patients, lawyers, and politicians, and it has become the main target in the discussions between researchers in the medical field, that of information related to health, and that of communication.

The analysis of the empirical data analysed demonstrated that the main elements motivating the Romanian doctors and students at the Faculty of Medicine's viewing of medical television series are medical knowledge, professional values, and team-work qualities.

The 'stories' told by the script of TV medical dramas were 'read' by the Romanian viewers as a narration about a set of universal values, which ensured their success. Thus, we witnessed the confirmation of the theory of 'uses and gratifications associated with the consumption of media messages' (McQuail, 1993, p. 73).

As other studies have showed, popular entertainment is increasingly acknowledged as a source of understanding about how social institutions, including health services, are structured (Storey, 1993; Tulloch, Lupton, 1997). The interviews made with medical staff and medical students had revealed the contextualizing role played by the consumption of medical TV series. Due to the comparisons made between Romanian hospitals and America ones (as they were presented in TV series), the respondents were able to assess both the fictional character of some parts of the script and the huge gaps between Romanian and other (mainly Western) medical systems.

As suggested by Vanderford (1999, p. 33):

> Television, as the dominant form of mass media, is an important source of shared perceptions which offers lessons for viewers about the way the world works.

In the case of our research project we can conclude that even though the respondents recognized the cultural artefacts as presenting either fictitious or heavily dramatized accounts of real incidents, nevertheless they were susceptible to adopting new definitions of reality, and to adjusting their value systems.

References

Babrow, A. S. (1987). Student motives for watching soap operas. *Journal of Broadcasting & Electronic Media*, 31, 309–321.

Barnet, E. (2007). *Dissecting the medical drama: A generic analysis of Grey's Anatomy and House*, MD. Department of Communication of Boston College. Retrieved from https://alumni.bc.edu/content/dam/files/schools/cas_sites/communication/pdf/thesis07.barnet.pdf

Brodie, M., Foehr, U., Rideout, V., Baer, N., Miller, C., Flournout, R., Altman, D. (2001). Communicating health information through the entertainment media. *Health Affairs*, 20(1), 192–199.

Cho, S. (2006). Network news coverage of breast cancer, 1974-2003. *Journalism and Mass Communication Quarterly*, 83(1), 116–130.

Czarny, M. J., Faden, R. R., Nolan, M. T., Bodensiek, E., Sugarman, J. (2008). Medical and nursing students' television viewing habits: Potential implications for bioethics. *American Journal of Bioethics*, 8(12), 1–8.

Davin, S. (2003). Healthy viewing: The reception of medical narratives. *Sociology of Health & Illness*, 25, 662–679.

Fishbein, M., Ajzen, I. (1975). *Belief, attitude, intention, and behavior: An introduction to theory and research*. Reading, MA: Addison-Wesley.

Flick, U. (1998). Everyday knowledge in social psychology. In U. Flick (Ed.) *The psychology of the social* (pp. 1–14). Cambridge: Cambridge University Press.

Freimuth, V.S., Stein, J.A., Kean, T.J. (1989). *Searching for health information: the cancer information service model.* Philadelphia, PA: University of Pennsylvania Press.

Gerbner, G., Morgan, M., Signorielli, N., Shanahan, J. (2002). Growing up with television: Cultivation processes. In J. Bryant, & D. Zillmann (Eds.) *Media effects: Advances in theory and research* (2nd ed). Hillsdale, NJ: Erlbaum.

Hether, H., Huang, G., Beck, V., Murphy, S. T., Valente, T. (2008). Entertainment-education in a media-saturated environment: Examining the impact of single and multiple exposures to breast cancer storylines on two popular medical dramas. *Journal of Health Communication*, 13, 808–823.

Hinyard, L. J., Kreuter, M. W. (2007). Using narrative communication asa tool for health behavior change: A conceptual, theoretical, and empirical overview. *Health Education & Behavior*, 34, 777–792.

Hornik, R., Niederdeppe, J. (2008). Information scanning. In W. Donsbach (Ed.) *International encyclopedia of communication* (pp. 2257–2261). Oxford, UK: Blackwell.

Hotz, R. L. (2002). The difficulty of finding impartial sources in science. *Nieman Reports*, 56, 6-7.

Joffe, H. (2003). Risk: From perception to social representation. *British Journal of Social Psychology*, 42, 55–73.

Johnson, J. D., Meischke, H. (1993). A comprehensive model of cancer-related information seeking applied to magazines. *Human Communication Research*, 19(3), 343–367.

Katz, E., Blumler, J. G., Gurevitch, M. (1974). The uses of mass communications: Current perspectives of gratifications research. In E. Katz, J. G. Blumler (Eds.) *Utilization of mass communication by the individual* (pp. 19–32). Thousand Oaks, CA: Sage.

Kruglanski, A. W. (2001). Social cognition, social representations, and the dilemmas of social theory construction. In K. Deaux, G. Philogene (Eds.) *Representation of the social* (pp. 242–248). Malden, MA: Blackwell Publishers.

Levy, M. R., Windahl, S. (1984). Audience activity and gratifications: A conceptual clarification and exploration. *Communication Research*, 11, 51–78.

Livingstone, S. M. (1988). Why people watch soap operas: An analysis of the explanations of British viewers. *European Journal of Communication*, 3: 55–80.

Lupton, D. (1994). *Moral threats and dangerous desires: AIDS in the news media.* Bristol, PA: Taylor & Francis, Inc.

McQuail, D. (1993). *Mass Communication Theory,* Sage Publications, London, Newbury Park, Beverly Hills, New Delhi.

Morgan, M. Signorielli, N. (1990). Cultivation analysis: Conceptualization and methodology. In N. Signorielli, M. Morgan (Eds.), *Cultivation analysis: New directions in media effects research.* Newbury Park, CA: Sage Publications, Inc.

Morgan, S. E., Movius, L., Cody, M. J. (2009). The power of narratives: The effect of entertainment television organ donation storylines on the attitudes, knowledge, and behaviors of donors and non-donors. *Journal of Communication*, 59(1), 135–151.

Moscovici, S. (1984). The phenomenon of social representations. In R. Farr &S. Moscovici (Eds.) *Social representations* (pp. 3–69). Cambridge: Cambridge University Press.

—. (1988). Notes towards a description of social representations. *European Journal of Social Psychology*, 18, 211–250.

Moscovici, S., Hewstone, M. (1983). Social representations and social explanations: From the 'naive' to the 'amateur' scientist. In M. Hewstone (Ed.) *Attribution theory* (pp. 48–125). Oxford: Blackwell.

Nabi, R. L., Biely, E. N., Morgan, S. J., Stitt, C. R. (2003). Reality based television programming and the psychology of its appeal. *Media Psychology*,5, 303–330.

Nelkin, D. (1995). *Selling science: How the press covers science and technology.* New York City: W.H. Freeman and Company.

Niederdeppe, J., Hornik, R. C., Kelly, B. J., Frosch, D. L., Romantan, A.,Stevens, R. S., Schwartz, J. S. (2007). Examining the dimensions of cancer-related information seeking and scanning behavior. *Health Communication*, 22, 153–167.

Palmgreen, P., Wenner, L. A., Rayburn, J. D. (1980). Relations between gratifications sought and obtained: A study of television news. *Communication Research*, 7, 161–192.

Papacharissi, Z., Mendelson, A. L. (2007). An exploratory study of reality appeal: Uses and gratifications of reality TV shows. *Journal of Broadcasting &*

Electronic Media, 51, 355–370.

Perse, E. (1990). Cultivation and involvement with local television news. In N. Signorielli, M. Morgan (Eds.) *Cultivation analysis: New directions in media effects research*. Newbury Park, CA: Sage Publications, Inc.

Perse, E. M. (1990). Media involvement and local news effects. *Journal of Broadcasting & Electronic Media*, 34, 17–36.

Petty, R. E., Cacioppo, J. T. (1986). The elaboration likelihood model of persuasion. In L. Berkowitz (Ed.) *Advances in experimental social psychology* (Vol. 19, pp. 123–205). New York: Academic Press.

Quick, B. L. (2009). The effects of viewing *Grey's Anatomy* on perceptions of doctors and patient satisfaction. *Journal of Broadcasting & Electronic Media*, 53, 38–55.

Rubin, A. M., Perse, E. M. (1987). Audience activity and soap opera involvement: A uses and effects investigation. *Human Communication Research*, 14: 246–268.

Shanahan, J., Morgan, M. (1999). *Television and its viewer: Cultivation theory and research*. Cambridge, UK: Cambridge University Press.

Shim, M., Kelly, B., Hornik, R. (2006). Cancer information scanning and seeking behavior is associated with knowledge, lifestyle choices, and screening. *Journal of Health Communication*, 11(Suppl. 1), 157–172.

Slater, M. D. (2002). Entertainment education and the persuasive impact of narratives. In M. C. Green, J. J. Strange, T. C. Brock (Eds.) *Narrative impact: Social and cognitive foundations* (pp. 157–181). Mahwah, NJ: Lawrence Erlbaum Associates.

Sommer, C. M. (1998). Social representations and media communications. In U. Flick (Ed.) *The psychology of the social* (pp. 186–195). Cambridge: Cambridge University Press.

Storey, J. (1993). *An introductory guide to cultural theory and popular culture*. Athens: The University of Georgia Press.

Tulloch, J., Lupton, D. (1997).*Television, AIDS, and risk: A cultural studies approach to health communication*. St. Leonards, Australia: Allyn & Unwin.

Vanderford, M. (1999). Television and religious values: A case study of ER and moral ambiguity. In F. Eigo (Ed.) *Religious values at the threshold of the third millennium* (pp. 33–73). Villanova, PA: Villanova University Press.

Vincent, R. C., Basil, M. D. (1997). College students' news gratifications, media use, and current events knowledge. *Journal of Broadcasting & Electronic Media*, 41, 380–392.

Voss, M. (2002). Checking the pulse: Mid-Western reporters' opinions on their ability to report health care news. *American Journal of Public Health,* 92, 1158–1160.

Wells J, Marshall P, Crawley B, Dickersin K. (2001). Newspaper reporting of screening mammography. *Annals of Internal Medicine*, 135(12), 1029–1037.

Wicclair, M. R. (2008). The pedagogical value of *House, M.D.*: Can a fictional unethical physician be used to teach ethics? *American Journal of Bioethics*, 8, 16–17.

Zukin, C., Snyder, R. (1984). Passive learning: When the media environment is the message. *Public Opinion Quarterly*, 48, 629–638.

Zurawik, D. (1999, January 27). Doctor shows often a prescription for success. *Lincoln Journal Star*, 4D.

CHAPTER EIGHT

'AMERICA'S FAVOURITE SERIAL KILLER': ENJOYMENT OF THE TV SERIAL *DEXTER*

DANIELA SCHLÜTZ, BEATE SCHNEIDER AND MAIK ZEHRFELD

The US-American programme *Dexter*[1] is an award-winning TV serial[2] (Showtime, 2006–2013). At the same time, it is a great success in terms of ratings, with the season 7 finale being the top-rated Showtime original programme in the channel's history so far (Kondolojy, 2012). As a consequence, the serial has been sold worldwide and is broadcasted in many European countries, Germany among them (since 2008). *Dexter* depicts the life and work of Dexter Morgan (Michael C. Hall)—according to Showtime, 'America's favourite serial killer'. Dexter is both a forensic expert (a blood-stain pattern analyst) with the Miami Metro Police Department and a serial killer. His double life is artfully visualized in the opening sequence of the programme (Karpovich, 2010). According to a specific code of conduct ('the code') given to him by his foster father, Dexter hunts down murderers who have escaped the justice system. Although Dexter only murders 'bad' people in order to satisfy his inherent urge to kill, the serial's narrative goes beyond the scope of the classical 'good guy–bad guy' scheme: The killer Dexter Morgan is depicted as a nice guy, a family man even. Thus, fans care about Dexter; they become his accomplices fearing for the killer to be caught.

Research on entertaining media content has shown the paramount importance of (positive) affective dispositions to feel enjoyment (Affective Disposition Theory, ADT; Zillmann & Cantor, 1976). ADT has received a

[1] To distinguish the program *Dexter* from its main character Dexter Morgan we will use the former term in italics.

[2] Judging by its narrative structure, *Dexter* is a continuous serial as opposed to an episodic series (cf. Cantor & Pingree, 1983; Creeber, 2004).

lot of empirical support during the last 30 years, with some disputed questions left: Why do people enjoy sad stories (Oliver, 1993a), expose themselves with delight to horror movies (Cantor, 2006; Oliver 1993b; 1994; Tamborini, 2003), cherish violent entertainment (Miron, 2003), or savour antihero narratives (Raney, Schmid, Niemann, & Ellensohn, 2009)? This paper deals with the latter puzzle: How do viewers establish a relationship with a fictional character who is morally imperfect, to say the least? Why do they feel entertained by a show that places an antihero in the centre of the narrative? Antiheroes are fictional characters that lack archetypical heroic features like bravery, chivalry and selflessness. They are more complex in terms of character. As Raney and Janicke (2013, p. 159) put it, 'we might compare traditional characters with a billiard ball: round, slick, smooth, shiny, and flawless. A morally complex character, in turn, is more like a golf ball: still round, but rough and notched.' Popular antiheroes can be found in literature (e.g. Lisbeth Salander in the 'Millennium Trilogy' by Stig Larsson) as well as in movies (e.g. Léon Montana in *Léon*) and TV series, for example Jack Bauer in *24*, Jackie Peyton in *Nurse Jackie*, Tony Soprano in *The Sopranos*, or, more recently, Dexter Morgan (*Dexter*) and Walter White (*Breaking Bad*). The worldwide popularity of these antiheroes suggests that they are capable of supporting a narrative that is enjoyable and entertaining—at least to some people.

Theoretical background

Scholars are puzzled by questions concerning affective dispositions toward a killer: 'How can we detest killing and organized crime but love *The Godfather* (and *The Godfather Part 2*)? How can we be sickened by senseless and random acts of crime but cheer for the lead characters in *Bonnie and Clyde*? How can we abhor theft but applaud the exploits of Danny Ocean and his gang in *Ocean's 11* and its sequels?' (Raney, 2011a, p. 171) In this paper, we wish to further explore how positive affective dispositions towards seemingly unworthy protagonists are established. In the following, we therefore discuss relevant findings from media entertainment research.

'Entertainment' is an open concept. Different viewers find pleasure in diverse media content. Whereas some like sport events, others prefer game shows; some play video games while others favour a good book; still others are fans of fictional TV serials like *Breaking Bad*, *The Wire* or *Dexter*. When asked what they derive from these activities, most of the fans would probably say that they are entertaining. Entertainment, then, might mean enjoyment, appreciation, suspense, fun, etc. Each user has different expectations and values different qualities in media entertainment. Fictional characters are central in explaining the effects of media entertainment, though. The athlete, the game show host, the avatar, the antihero or the lead in the latest TV series seems to play a key role in the enjoyment of media content. Hence, this is the focus of our paper.

Research on media entertainment has found affective dispositions a very important aspect of entertainment as media effect. ADT has proven fruitful to explain these effects (Raney, 2006) even under real-world conditions (cf. Weber, Tamborini, Lee, & Stipp, 2008). The theory 'posits that enjoyment of media content is a function of a viewer's affective disposition towards characters and the outcomes experienced by these characters in the unfolding narrative. Simply stated, it predicts that enjoyment will increase when liked characters experience positive outcomes and/or when disliked characters experience negative ones.' (Raney, 2006, p. 140) To be enjoyed, then, the narrative in question must allow for the character Dexter to be liked. Empathy ('feeling with', Zillmann, 2006a, p. 151) is central for emotional reactions to fictional characters. Intense empathy can lead to identification ('feeling as if', Zillmann, 2006a, p. 151; cf. Cohen, 2001) when the viewer adopts (at least partly) the persona's perspective, beliefs and emotions. Identification seems to be crucial for the enjoyment of antihero narratives whereas empathy is more important to the classical hero tale (Janicke & Raney, 2011; Raney & Janicke, 2013; Raney et al., 2009).

In *Dexter*, certain textual cues facilitate empathy. Voice-over technique, for instance, is used to offer insights into Dexter's character and his motivations. Empathy is furthermore supported by flashbacks showing Dexter's experiences. These techniques along with other aspects of the narrative (i.e., the code) are meant to make Dexter, the antihero, more likable. According to a content analysis of *Dexter* blogs and discussion forums in Germany and the USA, this strategy works quite well (Schlütz, Stock, Walkenbach, & Zehrfeld, 2013). Fans relate to Dexter in several ways: He is liked, even loved by some of the fans: 'I have completely fallen in love w/ the character Dexter and CAN NOT get enough,' somebody wrote on tv.com. Fans find him 'charming' and 'adorable'. They even identify with him: 'When I watched the first episode of Dexter, I said to myself, that is me to a 'T' without the killing,' one fan said on dexterwiki.sho.com. This leads to our first research question:

RQ1: Do viewers of Dexter *form affective dispositions toward the main character Dexter Morgan?*

Moral considerations are a vital part in developing affective dispositions

(Bryant & Miron, 2002, p. 568; Raney, 2002; 2004; 2005; 2011a; 2011b; Raney & Bryant, 2002). When we like a character, we take his or her side. This side-taking has to be morally justified. While watching a series, for instance, we continually observe the moral appropriateness of the character's actions and evaluate them accordingly. ADT holds that affective dispositions are formed more easily towards characters we evaluate as good because 'persons pursuing good causes (i.e., consensually approved causes) are considered good people, and […] good people are judged deserving of good fortunes. Analogously, persons pursuing bad causes (i.e., consensually condemned causes) are bad people, and bad people are judged deserving of bad fortunes; or, at the very least, undeserving of good fortunes' (Zillmann, 2006b, p. 229). Drama enjoyment, thus, is dependent (among other factors) on affective dispositions and moral judgment. Complex and morally ambiguous fictional characters such as Dexter may defy such categorizations (Raney & Janicke, 2013).

The morally ambiguous character of the antihero can be read differently, though. Dexter, for instance, only punishes people who 'deserve' it, like a paedophile who has killed several young boys (Father Donovan in the pilot of *Dexter*). This apparent justice makes Dexter the hero of the serial rather than the villain (Byers, 2010). *Dexter* fans seem to follow these narrative cues (at least in part). They give Dexter great moral license in order to keep up their joy (cf. Raney, 2011a). They rationalize moral considerations post hoc in order to cope with the moral dilemmas Dexter's actions pose (cf. Raney, 2004). The aforementioned content analysis of blogs and internet forums revealed these strategies (Schlütz et al., 2013). To be able to judge Dexter's actions as morally appropriate and thereby keep up the enjoyment, his deeds are legitimated. The justification strategies are threefold. Firstly, Dexter's upbringing is held responsible: because he has been traumatized in his youth, he must kill (one fan on tv.com writes: 'So, I see Dexter as another one of those victims, just trying to cope'). Secondly, the victims are vilified to satisfy Dexter's code: they are dubbed 'sub-human', 'filth', 'scum of the earth' or 'rats'. Therefore, they deserve to die: 'as long as he is killing the 'Human Trash' you should have nothing to worry about' (dexterwiki.sho.com). Thirdly, Dexter's deeds are either belittled ('He chops people into nice little pieces, now, I think that's a little monstrous', tv.com) or interpreted as useful, as 'serving the public interest' ('Dexter: the taxpayer's friend', dexterwiki.sho.com). These strategies reframe the meaning of Dexter's murders and rationalize his acts morally (cf. Hartmann & Vorderer, 2010). As a result, viewers are enabled to like Dexter in spite of what he does and, consequently, enjoy the show. The content analysis suggests that this is the case: the fictional character of Dexter Morgan is liked, even loved by some of the fans. They relate to Dexter and empathize with him although this is accompanied by mixed emotions and guilt. Thus, the second research questions reads as follows:

RQ2: Do viewers of Dexter *morally judge Dexter Morgan's actions?*

Hartmann and Vorderer (2010) indicated another important factor: They showed that familiarity with a violent computer game enhanced enjoyment because negative affects were weakened in the long run (p. 105; cf. Tamborini, Weber, Eden, Bowman, & Grizzard, 2010). Raney (2004) argued that viewers might develop story schema that influence their interpretations of the narrative. Schemas, then, function as an 'interpretive lens' (Raney & Janicke, 2013, p. 157): Frequent viewers 'develop media-related schema that include justification for immoral character actions which allow for the guilt-free enjoyment of reprehensible actions by likable characters' (Raney & Janicke, 2013, p. 163). The longer one is exposed to *Dexter*, the easier it becomes to justify Dexter's actions in order to keep up—or enhance—the enjoyment. This leads to research question three:

RQ3: Does familiarity with Dexter *influence moral judgment of Dexter?*

As moral considerations draw on values when judging someone as good or bad, the feeling of entertainment should be influenced by personal values. In a serial like *Dexter* with a challenging plot, individual differences in value systems should influence the feeling of entertainment. In the case of *Dexter*, this is especially true because crucial elements of the narrative are connected to specific values: *Dexter* 'draws upon and references the long-standing popularity of vigilante narratives in American pop culture [...] Fear of rising levels of violent crime, a perception that violent criminals routinely go uncaught and unpunished, and a more general level of skepticism about the effectiveness of the legal apparatus have all been long-standing features of American public discourse about law and order, and all of these factors play a role in the success of Dexter the series and the popularity of Dexter the character' (Schmid, 2010, p. 136-137). Raney and Bryant (2002) argue that in crime drama, audience inputs and message inputs interact to form a moral judgment. Viewers compare their notions of proper justice with the ones depicted in the show. The 'degree of correspondence between the viewer's sense of justice and the statement about the justice made in the drama' (Raney & Bryant, 2002, p. 407) has a bearing on the entertainment effect: The higher the congruity, the greater the enjoyment (cf. Schlütz, 2012). Thus, different moral beliefs should lead to divergent moral judgments when evaluating a given message input. This leads to research question number four:

RQ4: Do viewers' personal values have a bearing on the enjoyment of Dexter*?*

In sum, ADT models entertainment as a media effect grounded mainly in affective dispositions held towards deserving fictional characters. Whether characters are assessed as deserving or not is decided by morally judging the character's actions. Whether he or she is judged affirmatively, depends on individual values. When he or she is perceived as good, the viewer will fear for a positive outcome for him or her. If this is the case, the probability increases that the viewer will enjoy the show. At first sight, this should rule out a protagonist like the serial killer Dexter as suitable for entertainment. The programme's narrative, though, suggests specific justification strategies. Moreover, it offers cues for affective dispositions facilitating a positive evaluation of Dexter's actions. This evaluation may lead to empathy toward and identification with the serial killer. Hence, research question five reads:

RQ5: How do affective dispositions, moral evaluations, familiarity and individual values influence the enjoyment of Dexter*?*

Method

To address our research questions, we conducted an online survey in May 2010 among German *Dexter* fans. The questionnaire was distributed using a snowball system with several online starting points. 853 questionnaires were filled in completely. Due to size, sampling procedure and self-selection bias, the sample is not representative for the overall audience of *Dexter*.

We composed a questionnaire that dealt with the theoretical constructs discussed above. While some of the measures were customized, others were taken from validated scales.

Entertainment was measured on a 5-point Likert-type scale from 1 'does not apply at all' to 5 'applies completely' with 14 items. Following Trepte (2006), some of the items were taken from the Media Enjoyment Scale (MES) by Krcmar, others from the Reading Experience Scale by Appel, Koch, Schreier and Groeben (2002), still others were phrased individually to better fit our purpose (for all items see Tables 1–3). *Affective dispositions* were captured by means of an 18-item scale following Gleich (1997). To measure aspects of *moral judgment*, we phrased nine items with regard to the narrative.

To reduce the item lists and extract relevant dimensions, we conducted factor analyses (PCA, Oblimin and Varimax rotations, respectively) and extracted factors with eigenvalue > 1. Responses to all items per factor were then averaged for each participant to yield a single score. Higher index scores indicated higher agreement. All indices were tested for reliability. In some cases, items had to be excluded from the dimension to increase reliability (the results are shown in Tables 1–3).

Familiarity was operationalized by the number of seasons the respondents had already watched. At the time of the survey, four seasons had been aired (at least in the US). 58 percent had seen all of them (code 3), 26 percent up to three seasons (code 2) and 15.5 percent two seasons or less (code 1; $M = 2.4$, $SD = 0.7$).

To test for the influence of *individual values*, we phrased three items (scale from 1 'do not agree at all' to 5 'agree completely'). In connection with the *Dexter* narrative, they were meant to represent the notion of vigilantism (*Cronbach's Alpha* = .73): 'Some people deserve to die' ($M = 3.0$, $SD = 1.4$), 'In some cases vigilantism is right' ($M = 2.7$, $SD = 1.3$) and 'There is absolutely no justification for killing someone' (reversed; $M = 2.5$, $SD = 1.4$).

Our sample consisted of 853 respondents. 70.5 percent of the respondents were male, with an average age of 26 years ($M = 26.4$, $SD = 6.6$). The sample comprised mainly *Dexter* fans: 60 percent of the respondents named *Dexter* as one of their top-three favourite programmes in an open-ended question. 83 percent agreed completely with the item 'I like watching *Dexter*', men and women alike. As fans and nonfans of morally complex characters differ in terms of their entertainment experience (Janicke & Raney, 2011) this confinement is reasonable to achieve a clearer pattern.

Results

A premise of our arguing is that the TV serial *Dexter* is entertaining. This is indeed the case, as was shown above. We now want to delve deeper into the entertainment experience of *Dexter*. Therefore, we performed a Principal Component Analysis (PCA) with Oblimin Rotation[3] with the 14-item scale. Two dimensions were extracted: enjoyment ($M = 4.7$, $SD = 0.5$) with items that comprised an affective reaction to the show, and involvement covering cognitive reactions ($M = 3.5$, $SD = 0.8$; cf. Table 1). Respondents reported a very high level of enjoyment of the show.

---

[3] We chose an oblique rotation method in order to account for the fact that the factors are correlated. Therefore, explained variance cannot be computed.

Involvement was considerably lower, though.

Table 8-1. Dimensions of entertainment from *Dexter*: Enjoyment and involvement (N = 853)

		Dimensions	
	Item Mean*	Enjoyment	Involvement
(1) I like watching *Dexter*.	4.8	0.810	
(2) *Dexter* is very entertaining.	4.6	0.785	
(3) *Dexter* is an exciting series.	4.7	0.769	
(4) *Dexter* has a good plot.	4.6	0.703	
(5) I always look forward to seeing the next episode of *Dexter*.	4.6	0.678	
(6) Sometimes I find *Dexter* too bloody/brutal.**	1.6	-0.482	
(7) Even when I am not watching *Dexter* I think about it.	2.9		0.810
(8) Without *Dexter* I would miss out on something.	3.1		0.695
(9) There are some episodes of *Dexter* I have already watched several times.	2.9		0.666
(10) *Dexter* is really special to me.	3.7		0.629
(11) I seek additional information about *Dexter* in other media (e.g., Internet).	3.4		0.613
(12) I think about what could happen next in *Dexter*.	3.9		0.611
(13) When watching *Dexter* I do not want to be disturbed.	4.1		0.441
(14) There should be more shows like *Dexter*.***	3.9		
Cronbach's Alpha (for Index)		.87	.79
Factor Mean (SD)		4.6(0.5)	3.4(0.8)

Question: „*Let us talk about Dexter. What do you think about the series in general? Please specify how much the statements apply to you.*"
Note: PCA, Oblimin Rotation
*5-point Likert-type scale from (1) 'does not apply at all' to (5) 'applies completely'
**Item was removed from the scale to enhance internal reliability.
***Item was not considered for average score.

RQ1 was concerned with the main character's ability to generate affective dispositions (Do viewers of *Dexter* form affective dispositions toward the main character Dexter Morgan?). Again, we performed a PCA with Oblimin Rotation. We found three dimensions of affective dispositions (cf. Table 2): identification ($M = 2.7$, $SD = 0.8$), empathy ($M = 4.0$, $SD = 0.7$) and antipathy ($M = 2.3$, $SD = 0.7$). Apparently, respondents felt empathy with Dexter, but they rarely identified with him. This is in line with findings of Janicke and Raney (2011). In their study on Jack Bauer in *24*, identification was not statistically significant in predicting fan enjoyment as well. In the case of *Dexter*, the reported feeling of antipathy is very low. Apparently, viewers empathize with Dexter Morgan. They do not identify with the killer, though. All three dimensions correlated with the two entertainment dimensions according to theoretical assumptions: Identification was positively correlated with both involvement ($r = .418$, $p \leq .001$) and enjoyment ($r = .222$, $p \leq .001$) as was empathy ($r_{inv} = .548$, $p \leq .001$; $r_{enj} = .528$, $p \leq .001$). Antipathy, on the other hand, was associated negatively with enjoyment ($r = -.378$, $p \leq .001$) and involvement ($r = -.314$, $p \leq .001$).

Table 8-2. Dimensions of affective dispositions toward *Dexter*: Identification, empathy, and antipathy (N = 853)

	Item Mean*	Dimensions		
		Ident.	Emp.	Antip.
(1) Sometimes I think that I want to be like Dexter.	1.9	0.786		
(2) Sometimes I feel similar to Dexter.	2.4	0.746		
(3) I would like to be friends with Dexter.	2.7	0.657		
(4) Often, I feel spoken to directly by Dexter.	2.5	0.639		
(5) I would like to meet Dexter in person.	2.9	0.562		
(6) I trust Dexter.	2.9	0.489		
(7) Dexter possesses certain personal qualities that I respect and admire him for.***	3.6	0.444		

	Item Mean*	Dimensions (*cont.*)		
		Ident.	**Emp.**	**Antip.**
(8) Sometimes, I feel compassion for Dexter.	4.0		0.783	
(9) Dexter often expresses thoughts that make me think.	3.6		0.558	
(10) Since I can hear Dexter's thoughts I feel connected with him.	3.5		0.547	
(11) I always look forward to seeing Dexter again.	4.2		0.542	
(12) I find Dexter attractive.**	2.9		0.466	
(13) Sometimes I have problems liking Dexter.	2.3			0.799
(14) Some of Dexter's actions or utterances I cannot comprehend at all.	2.6			0.672
(15) I like Dexter Morgan.	4.2		0.474	-0.523
(16) I can easily overlook Dexter's mistakes.	2.9			-0.461
(17) Dexter is nothing but a psychopathic monster.	2.0			0.431
(18) Most of the time I can understand Dexter quite well.**	3.6			
Cronbach's Alpha (for Index)		.83	.64	.65
Factor Mean (SD)		2.7 (0.8)	4.0 (0.7)	2.3 (0.7)

Question: *"A few more statements about Dexter Morgan. Please consider how much they apply to you."*
Note: PCA, Oblimin Rotation
*5-point Likert-type scale from (1) 'does not apply at all' to (5) 'applies completely'
**Item was removed from the scale to enhance internal reliability.
***Item was not considered for average score.

Regarding research question two (Do viewers of *Dexter* morally judge Dexter Morgan's actions?), we performed a PCA with Varimax Rotation.[4] We found two dimensions (cf. Table 3): justification ($M = 2.7$, $SD = 0.9$) and condemnation ($M = 3.3$, $SD = 0.9$). The two factors accounted for 50

[4] In this case, we chose the orthogonal rotation method because the dimensions are presumably independent from each other.

percent of the variance.

Table 8-3. Dimensions of morally judging Dexter (N = 853)

	Item Mean*	Dimensions Justifi-cation	Condem-nation
(1) Actually, Dexter is preserving lives by preventing murderers from killing again.	3.6	0.753	
(2) As long as Dexter keeps following the code everything is fine.	3.0	0.752	
(3) Dexter's murders are OK, as he is only killing bad people.	2.6	0.747	
(4) Sometimes I am glad for Dexter when he kills someone.**	3.0	0.549	
(5) There should be real people doing the things Dexter does.	1.8	0.522	-0.415
(6) I know that I should not side with Dexter.**	3.8		0.708
(7) What Dexter does is illegal and deserves to be punished.	3.3	-0.413	0.616
(8) If Dexter were real I would betray him.	2.8	-0.436	0.589
(9) Dexter does not kill to help; he does it because of his inner desire to kill.	3.8		0.581
Explained Variance (%)		29.4	21.0
Cronbach's Alpha (for Index)		.76	.62
Factor Mean (SD)		2.7(0.9)	3.3(0.9)

Question: *'The next part is about Dexter Morgan's actions. How much do the statements listed below apply to you?'*
Note: PCA, Varimax Rotation
*5-point Likert-type scale from (1) 'does not apply at all' to (5) 'applies completely'
**Item was removed from the scale to enhance internal reliability.

Apparently, the assessment of Dexter as acting in a morally condemnable way is more prevalent with the respondents than the justification of his actions. The respondents agreed that Dexter acts fairly and preserves lives by murdering murderers. That did not justify the killings completely, though. Respondents especially disagreed with the notion that the fictional character should be transferred into real life. Apparently, they draw a distinct line between fiction and reality.

The justification and condemnation of Dexter's actions was associated with the reported affective dispositions. The more Dexter and his actions were justified, the more the respondents identified ($r = .494, p \leq .001$) and empathized ($r = .290, p \leq .001$) with him. Antipathy, on the other hand, was negatively correlated with justification ($r = -.369, p \leq .001$). Congruent with Affective Disposition Theory, we observed a negative correlation between the condemnation of Dexter's deeds and both empathy ($r = -.181, p \leq .001$) and identification ($r = -.446, p \leq .001$), whereas antipathy was positively correlated ($r = .372, p \leq .001$).

The third question asked whether familiarity with *Dexter* influences moral judgment of Dexter. A correlation analysis shows that this is not the case. Neither justification nor condemnation are significantly associated with familiarity.

Regarding the fourth question (RQ4: Do viewers' personal values have a bearing on the enjoyment of *Dexter*?), we correlated the vigilantism index with the enjoyment index (while controlling for age and gender) and found a significant but rather small association ($r = .082, p \leq .05$). Apparently, real life values have only a very small bearing on fictional content—at least in our sample.

Finally, we conducted a stepwise regression analysis on enjoyment to account for RQ5 (How do affective dispositions, moral evaluations, familiarity and individual values influence the enjoyment of *Dexter*?). We chose enjoyment as independent variable because the affective reactions to the show that are contained in this dimension are closest to the entertainment experience (Vorderer, Klimmt, & Ritterfeld, 2004). We fitted all relevant scores in the model as independent variables. The results are depicted in Table 4. The independent variables explained 41 percent of the variance of the dependent variable. The model shows that involvement has by far the strongest effect on enjoyment of *Dexter*. The second most important factor is empathy followed by identification. Interestingly, identification has a negative influence. Respondents who identify more strongly with Dexter seem to feel less enjoyment. Condemnation, on the other hand, is positively correlated which means that respondents who condemn Dexter to a stronger degree report more enjoyment. The same holds true for justification. The latter is in line with the theory, though.

Table 8-4. Regression model ($R^2_{adj.}$ = .409) with enjoyment as dependent variable (N = 853)

	B	Std. Error	Beta	T	Sig.
(Constant)	3.217	0.142		22.726	.000
Involvement	0.216	0.022	0.331	9.941	.000
Empathy	0.222	0.028	0.290	7.960	.000
Familiarity	0.116	0.019	0.177	6.253	.000
Antipathy	-0.108	0.025	-0.141	-4.349	.000
Identification	-0.083	0.021	-0.131	-4.045	.000
Variables excluded from the model: condemnation, justification, vigilantism, gender and age					

Discussion

Our study dealt with an antihero's capability to entertain. We conducted an online survey to ascertain whether the fictional serial killer Dexter Morgan is entertaining in spite of his moral ambiguity. As ADT suggests, the development of affective dispositions towards good characters is vital for a feeling of entertainment. Our findings suggest that the serial *Dexter* is indeed perceived as entertaining. Especially affective aspects—or enjoyment—are 'at the heart of entertainment' of *Dexter* (Vorderer, Klimmt, & Ritterfeld, 2004). Involvement, i.e. the cognitive occupation with the serial beyond the reception process, is reported more rarely. Emotional reactions seem to take precedence over a cognitive analysis of the moral dilemma at stake. A thorough cognitive examination of the programme and its inherent dilemma fuels enjoyment, though.

Apparently, an antihero show like *Dexter* is enjoyed. Why is that? The respondents report positive affective dispositions toward Dexter, mainly empathy (compassion and a feeling of connectedness). Identification with Dexter, on the other hand, is referred to less often. Viewers trust Dexter, they would like to meet and befriend him, but they do not want to be like Dexter. This runs counter to Janicke and Raney's (2011) suggestion that the key to understanding antihero enjoyment is identification.

In line with ADT, empathy and identification correlate strongly with enjoyment and involvement. Antipathy, on the other hand, is negatively associated with enjoyment and involvement, which fits the theory as well. Moral evaluation, too, plays a role in evaluating *Dexter*. Respondents condemn Dexter's actions rather than justify them. They especially distinguish between fiction and reality. Again according to ADT, justification of Dexter's actions correlates positively with positive affective dispositions while condemnation is negatively associated.

Thus, all bivariate correlations support ADT. The regression model confirms these results to a large extent: Involvement and empathy enhance enjoyment, as does familiarity. However, as familiarity is not associated with moral justifications, this might not be due to the development of an 'interpretive lens' (Raney & Janicke, 2013, p. 157) that facilitates moral judgment. Maybe the reason lies within the narrative of *Dexter*, which evolves from season to season. This question has to remain open.

Other results of the regression analysis, however, aberrate from a classic ADT pattern. More identification, for instance, leads to less enjoyment (once the influence of involvement and empathy is accounted for). This conflicts with the results of Raney et al. (2009) and Janicke and Raney (2011). In the case of *Dexter*, viewers seemingly care for the protagonist without wanting to be like him. Additionally, moral judgment (condemnation and justification) does not play a significant role in explaining enjoyment in our study. This is in line with the findings of Janicke and Raney (2011). Liking seems to be the key in understanding enjoyment of antihero narratives. If you care for the immoral protagonist, you are prone to forgive him his moral shortcomings. This reduces cognitive dissonance and helps you to continue enjoying the show (cf. Raney 2011a; Raney & Janicke, 2013). However, as opposed to the argument brought forward by Raney and Janicke (2013), emotional connections with Dexter are not established by identifying with him.

Personal values like the attitude towards vigilantism do not have a bearing on the enjoyment of *Dexter* (after accounting for classic ADT factors)—at least not in our sample of fans.

Limitations

Despite the interesting and cohesive findings, there are, of course, limitations to our study. The first is related to the sample: Due to sampling size, sampling process, and self-selection bias, the sample is not representative for the audience of *Dexter*. However, as the respondents were frequent viewers of the series, their answers are meaningful. To measure realistic effects and thereby enhancing external validity, we wanted to question people who had had a chance to get familiar with *Dexter* and establish a relationship with the protagonist (cf. Raney et al., 2009). Of course, the generalizability of the data is limited with regard to other antihero series. Building affective dispositions with Dexter might be somewhat different from doing the same with other antiheroes like Walter

White. Secondly, our results cannot account for causal relationships due to our research design. What we can conclude, though, is that there are associations between affective dispositions, moral evaluations and enjoyment in the case of *Dexter*. Thirdly, our explicit, post-hoc survey design may have fostered rationalizations. Further studies should try to capture evaluation during the reception process, at best using implicit measures (cf. Hefner, Rothmund, Klimmt, & Gollwitzer 2011). This would provide an opportunity to ascertain whether moral judgment precedes liking or the other way around (Raney, 2004).

Conclusion

The TV serial *Dexter* was very well suited to our overall research question concerning the entertainment potential of antihero narratives. At the serial's heart is a moral dilemma. This dilemma is mirrored in the respondents' reactions to the drama and its main character. Our study was able to add to entertainment research in the field of ADT. Antihero narratives (at least the one we did research on) are perceived more or less like classic entertainment hero stories. The theory proved fruitful in explaining 'real life' serial enjoyment. Further research should pursue this avenue and test ADT and its additions with regard to existing media content. Especially the process of moral disengagement should be explored with reference to other antihero narratives (like *Breaking Bad*; cf. Janicke & Raney, 2012). Dependent on the justification strategies (either set up in the narrative or used by viewers), different effects might occur. Genuine fans should be explored within a longitudinal design to account for long-term effects that follow from prolonged exposure to specific media content (cf. Tamborini et al., 2010; Weber et al., 2008). With the proliferation of morally dubious but nevertheless (or precisely because of that) very successful TV series dealing with sex, violence or drugs (like *Californication*, *Dexter*, *Breaking Bad*, *The Wire*, *True Blood* or *Boardwalk Empire*), this could be an interesting and socially rewarding area of research.

References

Appel, M., Koch, E., Schreier, M., & Groeben, N. (2002). Aspekte des Leseerlebens: Skalenentwicklung [Aspects of the reading experience: Scale development]. *Zeitschrift für Medienpsychologie, 14*(4), 149–154. DOI: 10.1026//1617-6383.14.4.149

Bryant, J., & Miron, D. (2002). Entertainment as media effect. In J. Bryant, & D. Zillmann (Eds.), *Media effects: Advances in theory and research* (2nd. ed., pp.

549–582). Mahwah, NJ: Lawrence Erlbaum Associates.
Byers, M. (2010). Neoliberal Dexter? In D. L. Howard (Ed.), *DEXTER: Investigating cutting edge television* (pp. 143–156.). London: I. B. Tauris.
Cantor, J. (2006). Why horror doesn't die: The enduring and paradoxical effects of frightening entertainment. In J. Bryant, & P. Vorderer (Eds.), *Psychology of entertainment* (pp. 315–327). Mahwah, NJ: Lawrence Erlbaum Associates.
Cantor, M. G., & Pingree, S. (1983). *The soap opera*. Beverly Hills, CA: Sage.
Cohen, J. (2001). Defining identification: A theoretical look at identification of audiences with media characters. *Mass Communication & Society, 4*, 245–264.
Creeber, G. (2004). *Serial television: Big drama on the small screen*. London: BFI.
Cronan, M. (2009). *CCI: Dexter*. Retri eved from http://www.comicbookresources.com/ ?page=article&id=22284
Gleich, U. (1997). Parasocial interaction with people on the screen. In R. Winterhoff-Spurk, & T. H. A. van der Voort (Eds.), *New horizons in media psychology: Research cooperation and projects in Europe* (pp. 35–55). Olpaden: Westdeutscher Verlag.
Hartmann, T. & Vorderer, P. (2010). It's okay to shoot a character: Moral disengagement in violent video games. *Journal of Communication, 60*, 94–119.
Hefner, D., Rothmund, T., Klimmt, C., & Gollwitzer, M. (2011). Implicit measures and media effects research: Challenges and opportunities. *Communication Methods and Measures, 5*(3), 181–202.
Janicke, S. H., & Raney, A. A. (2012, May). *Modeling the antihero narrative enjoyment process*. Paper presented at the Annual Meeting of the International Communication Association, Phoenix, AZ.
Janicke, S. H., & Podwalski, P. (2012, November). *Differences in antihero enjoyment between Germany and the US.* Paper presented at the Annual Meeting of the National Communication Association, Orlando, FL.
Janicke, S. H., & Raney, A. A. (2011, May). *Exploring how we enjoy antihero narratives: A comparison of fans and nonfans of 24*. Paper presented at the Annual Meeting of the International Communication Association, Boston, MA.
Karpovich, A. I. (2010). Dissecting the opening sequence. In D. L. Howard (Ed.), *DEXTER: Investigating cutting edge television* (pp. 27–42). London: I. B. Tauris.
Kondolojy, A. (December 17, 2012). Dexter *and* Homeland *season finales deliver both series' highest-rated nights ever.* Online. R Retrieved from http://tvbythenumbers.zap2it.com/2012/12/17/dexter-and-homeland-season-finales-deliver-both-series-highest-rated-nights-ever/162148/.
Miron, D. (2003). Enjoyment of violence. In J. Bryant & J. Cantor (Hrsg.), *Communication and emotion: Essays in honor of Dolf Zillmann* (S. 445–472). Mahwah, NJ: Lawrence Erlbaum Associates.
Oatley, K. (1994). A taxonomy of the emotions of literary response and a theory of identification in fictional narrative. *Poetics, 23*, 53–74.
Oliver, M. B. (1993a). Exploring the paradox of the enjoyment of sad films. *Human Communication Research, 19*(3), 315–342.
—. (1993b). Adolescents' enjoyment of graphic horror: Effects of viewers'

attitudes and portrayals of victim. *Communication Research, 20*(1), 30–50.

—. (1994). Contributions of sexual portrayals to viewers' responses to graphic horror. *Journal of Broadcasting & Electronic Media, 38*(1), 1–17.

Raney, A. A. (2002). Moral judgment as a predictor of enjoyment in crime drama. *Media Psychology, 4*, 305–322.

—. (2004). Expanding disposition theory: Reconsidering character liking, moral evaluations, and enjoyment. *Communication Theory, 14*, 348–369.

—. (2005). Punishing media criminals and moral judgment: The impact on enjoyment. *Media Psychology, 7*, 145–163.

—. (2006). The psychology of disposition-based theories of media enjoyment. In J. Bryant, & P. Vorderer (Eds.), *Psychology of entertainment* (pp. 137–150). Mahwah, NJ: Lawrence Erlbaum Associates.

—. (2011a). Media enjoyment as a function of affective dispositions toward and moral judgment of characters. In K. Doeveling, C. von Scheve, & E. A. Konijn (Eds.), *The Routledge handbook of emotions and mass media* (pp. 166–178). London: Routledge.

—. (2011b). The role of morality in emotional reactions to and enjoyment of media entertainment. *Journal of Media Psychology, 23*(1), 18–23.

Raney, A. A., & Bryant, J. (2002). Moral judgement and crime drama: An integrated theory of enjoyment. *Journal of Communication, 52*, 402–415.

Raney, A. A., & Janicke, S. H. (2013). How we enjoy and why we seek out morally complex characters in media entertainment. In R. Tamborini (Ed.), *Media and the moral mind* (pp. 152–169). London: Routledge.

Raney, A. A., Schmid, H., Niemann, J., & Ellensohn, M. (2009, May). *Testing affective disposition theory: A comparison of enjoyment of hero and antihero narratives*. Paper presented at the Annual Meeting of the International Communication Association, Chicago, IL.

Schlütz, D. (2012). Der Prozess grenzüberschreitender Medienwirkungen: Das Susceptibility to Imported Media (SIM)-Modell am Beispiel US-amerikanischer Fernsehserien [Modelling cross-border media effects: The 'Susceptibility to Imported Media' (SIM) model and US-American TV series]. In H. Wessler, & S. Averbeck-Lietz (Eds.), *Grenzüberschreitende Medienkommunikation* [Transborder media communication] (pp. 183–202). Baden-Baden: Nomos.

Schlütz, D., Stock, Y., Walkenbach, J., & Zehrfeld, M. (2013). Mein Freund, der Serienkiller: Fan-Beziehungen zum Hauptcharakter der Fernsehserie Dexter [My friend, the serial killer: Fan-relations with the main character of the TV series *Dexter*]. In S. Eichner, L. Mikos, & R. Winter (Eds.), *Transnationale Serienkultur: Theorie, Ästhetik, Narration und Rezeption neuer Fernsehserien* [Transnational serial culture: Theory, aesthetics, narration and reception of new TV series] (pp. 367–380). Wiesbaden: VS.

Schmid, D. (2010). The devil you know: Dexter and the 'goodness' of American serial killing. In D. L. Howard (Ed.), *DEXTER: Investigating cutting edge television* (pp. 132–142). London: I. B. Tauris.

Tamborini, R. (2003). Enjoyment and social functions of horror. In J. Bryant, D. Roskos-Ewoldson, & J. Cantor (Eds.), *Communication and emotion: Essays in honor of Dolf Zillmann* (pp. 417–443). Mahwah, NJ: Lawrence Erlbaum Associates.

Tamborini, R., Weber, R., Eden, A., Bowman, N. D., & Grizzard, M. (2010). Repeated exposure to daytime soap opera and shifts in moral judgment toward social convention. *Journal of Broadcasting & Electronic Media, 54*, 621–640.

Tinchev, V. (2010). Dexter's German reception: Why are German networks so obsessed (and troubled) with US shows? In D. L. Howard (Ed.), *DEXTER: Investigating cutting edge television* (pp. 157–171). London: I.B. Tauris.

Trepte, S. (2006). Die Messung des Unterhaltungserlebens [Measuring the entertainment experience]. In W. Wirth, H. Schramm, & V. Gehrau (Eds.), *Unterhaltung durch Medien: Theorie und Messung* [Media entertainment: Theory and measures] (pp. 142–173). Cologne: von Halem.

Vorderer, P., Klimmt, C., & Ritterfeld, U. (2004). Enjoyment: At the heart of media entertainment. *Communication Theory, 14*(4), 388–408

Weber, R., Tamborini, R., Lee, H. E. & Stipp, H. (2008). Soap opera exposure and enjoyment: A longitudinal test of disposition theory. *Media Psychology, 11*(4), 462–487.

Zillmann, D. (2006a). Empathy: Affective reactivity to others' emotional experiences. In J. Bryant, & P. Vorderer (Eds.), *Psychology of entertainment* (pp. 151–182). Mahwah, NJ: Lawrence Erlbaum Associates.

—. (2006b). Dramaturgy for emotions from fictional narration. In J. Bryant, & P. Vorderer (Eds.), *Psychology of entertainment* (pp. 215–238). Mahwah, NJ: Lawrence Erlbaum Associates.

Zillmann, D., & Cantor, J. R. (1976). A disposition theory of humor and mirth. In A. J. Chapman, & H. C. Foot (Eds.), *Humor and laughter: Theory, research and applications* (pp. 93–115). London: Wiley.

CHAPTER NINE

AUDIENCE PERCEPTIONS OF HEALTH IN *GREY'S ANATOMY* TV SERIES

BIANCA MITU

Introduction

Who has health, has hope;
and who has hope, has everything
(Thomas Carlyle)

The media can influence people's perception about a particular topic or it can change attitudes, feelings or behaviours. The media has become a scapegoat for all sorts of societal and cultural ills. Critics blame the media for everything from obesity to the murder rate. Health is a contested concept which has been defined in various ways. The ways in which health is defined or understood can have wide-ranging implications and can impact upon issues such as health promotion or health education. As Jones points out, 'definitions of health contain within them complex ideas about what it means to be healthy, whose responsibility it is to maintain and how illness and disease should be interpreted' (Jones, 2000, p. 18). Health can also be considered as a socially constructed phenomenon, as Curtis (2004) suggests: health is related to individual and collective ideas and beliefs about identity, the nature of the body and its significance. While a growing body of scholarship has engaged with the issue of health, there is a lack of empirical research linking people's health beliefs and behaviours to the media, and in particular to their evaluations of health information provided by the media. Television has the strongest influence on the general public; as Jacobs also asserts, 'television fiction has the potential to amplify and refine the anxieties, hopes and despair of culture and society. We often discover through television dramas structures of feeling (and anxiety), ways of thinking and modes of behaviour that we recognise as congruent or adjacent with our own or with how we image other lives might be. At the same time television, like any art form, develops its own rules, conventions and ways of thinking in relation to its past. This is true for television fiction, particularly that set in the contemporary world, where the requirements of plausibility are often in tension with the necessity for dramatic invention, and the opportunities and constraints of genre and formula' (2003, p. 149). Like Jacobs, many other scholars draw attention to the fact that television has undoubtedly changed the way contemporary society functions and the way people view themselves. Television is an absorbing and fascinating medium. That is why people's ability to discern between facts and fiction is often blurred. Television's impact on individuals, communities, and cultures is huge. The television audience cannot help but internalize what it sees everyday on the TV screen. This also happens with health issues, health practices and behaviours that people observe in various TV shows or medical dramas. When investigating people's health beliefs and behaviours the medical drama genre needs special attention because although it can be entertaining it is also very important for its potential to inform people and persuade them regarding various health issues. The aim of the present study is to provide a comprehensive analysis of the relationship between TV series and the construction of people's health beliefs and behaviours, to discover the most viewed medical dramas, and to investigate Romanian people's reasons for watching medical dramas. We choose to focus on the TV series *Grey's Anatomy* because since its launch in 2005, *Grey's Anatomy* has been at the top of people's preferences with an audience in the range of 20 million viewers. Also, *Grey's Anatomy* is a TV series that includes 'an abundance of medical situations encountered by the hospital's staff. There are a multitude of opportunities for communicating health information to the public—whether as part of a conscious effort at education, or incidentally, as part of a naturally occurring storyline' (Rideout, 2008, p. 1).

Understanding Health

The concept of health is a contested concept which in the past decades has been defined in various ways. Health is becoming an important domain for all modern societies mainly because the health of people in one country is closely dependent on the health of people elsewhere. In the context of globalization, health threats are rapidly spread across borders. As well as the concept of health, globalization has been highly disputed in the past decades. Nowadays scholars have devoted more and more consideration to the widely discussed idea of identity formation in the age of globalization. Globalization is a controversial issue of considerable debate and constant disagreement. One of the main concerns of the thesis of globalization is the constant changing of time-space relations. This phenomenon has serious consequences and a great impact upon people's lives and culture, being related to how people perceive themselves and how people relate to others. In the context of an increasing phenomenon of globalization health has become an important issue on national and international political agendas.

The academic literature relating to the understanding of the concept of health is generally dominated by two major contrasting understandings of health: the biomedical model and the social model. The biomedical model is well-known to be the basis for western medicine, and therefore it has been extremely influential in most of the developed countries around the world. The biomedical model draws heavily on biology and other clinical sciences and offers physical, biological, or chemical explanations for the concept of health and ill-health. This particular model of health has often been linked to the idea of the 'body as a machine'. Also, according to the biomedical model, illness is considered to be a physical malfunction of the human body or of one of its components, which requires either correction or some form of treatment to reduce further malfunction. In contrast to illness, health is seen largely as the absence of disease (Curtis, Taket, 1996; Jones, 2000; Lloyd, Shakespeare, 2000). In contrast, the social model of health considers health to be a condition which is influenced by many social, political, economic, cultural, psychological, and environmental factors as well as biological ones (Jones, 2000).

These two models of understanding health, the biomedical model and the social model, are understandings of health which have been developed by both scholars and medical practitioners. Nevertheless, elements of these models are apparent in many lay understandings of health among ordinary people. Curtis (2004) suggests the main lay understandings of health are the ideas of health as balance and ill-health as imbalance, the body as a machine and illness as a malfunction of this machine, health as a locus of control (the individual must have control of health), health as the outcome of fate or divine will, health as a freedom to do as one pleases or carry out the key roles required, health as a resilience against infection or hazards, and health as the access to the means for good health. Over time, scholars have paid attention to ordinary people's understanding of health as well. The first three main studies related to people's understanding of health are Blaxter and Paterson (1982), Williams (1983), and Calnan (1987). Based on different samples these studies identify different concepts of health. Blaxter and Paterson's conceptions of health are: health as the absence of disease or illness requiring hospitalization; being able to work and carry out normal daily tasks; the capacity to cope with illness, pain, or disability; and, of course, health as a strength of character and a matter of willpower and self-discipline. According to this study, general well-being and physical fitness did not feature in respondents' conceptions of health but we can notice them in other studies. Williams's study related to the understanding of health identifies two dominant conceptions of health: health seen as the absence of illness and disease and health as strength. The main conceptions identified in Calnan's study about ordinary people's understanding of health are: health as the absence of illness; health as an ability to cope with life crises, including illness and chronic illness; health as being fit, strong, active and taking exercise; and health as meeting the demands of daily living and being able to get through the day without problems. Calnan also stated that it was important to look at the differences between social classes, because people from different social classes usually have different concepts of health. Anyway, as Curtis notes individuals often work with more than one framework or understanding of health, creating 'complex and variable frames of reference' (Curtis, 2004, p. 3).

Trying to approach the relation between communication and health it is easy to assume that communication bears a great importance for people's health because it can:

'- increase knowledge and awareness of a health issue, problem, or solution
- influence perceptions, beliefs, attitudes, and social norms
- prompt action
- demonstrate or illustrate skills
- show the benefit of behavior change
- increase demand for health services
- reinforce knowledge, attitudes, and behaviors
- refute myths and misconceptions
- help coalesce organizational relationships
- advocate for a health issue or a population group' (Thomas, 2005, p. 4).

Television is a great means of communicating health beliefs and behaviours. Since its foundation, television has received a crucial role—to be in the public service—and that means it has obligations regarding the education and informing of citizens. People's minds are connected through shared-emotions, shared-experiences, shared-history, and shared-stories, and television is the only medium able to connect a large amount of people.

Romanian people and medical TV Series

Starting from these considerations, the present study aims to find out how medical TV series can influence people's perceptions of health. Furthermore, the study focuses on the most viewed medical TV series and investigates Romanian people's reasons for watching medical dramas, in particular *Grey's Anatomy*. The survey was conducted in February 2014 and used a sample of 500 regular TV series viewers of all ages, Romanian people who indicate that they usually watch medical TV series. Only 430 responses were valid for the present study, 41% men and 59% women.

According to the results of the survey, Romanian people spend several hours everyday watching TV series (Figure 1).

Figure 9-1 - The time people spend watching TV series

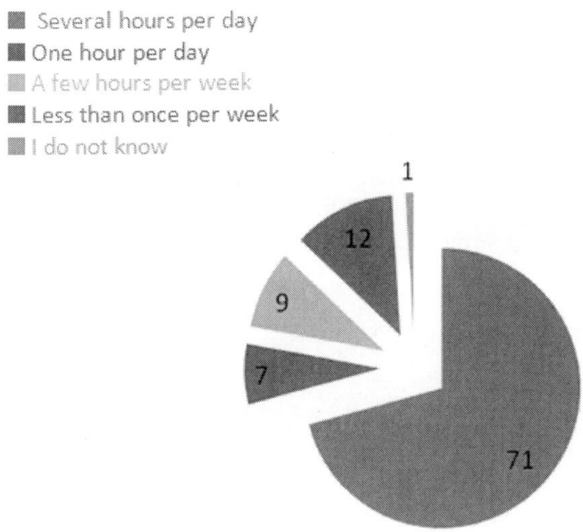

An important aspect of the survey was to find out the most viewed medical TV series in Romania. According to the results of the survey, Romanian people choose *Grey's Anatomy* (73%), *House M.D.* (18%) and *Private Practice* (8%).

Figure 9-2. Most viewed medical TV series

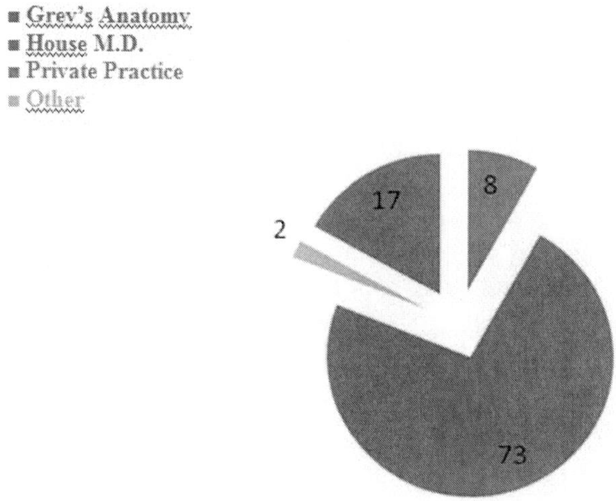

The main reasons for watching *Grey's Anatomy* are the interpersonal relationships such as friendship, love relationships, professional relationships (64%), gathering information about health problems (21%) and curiosity (15%).

Figure 9-3. Reasons for watching *Grey's Anatomy*

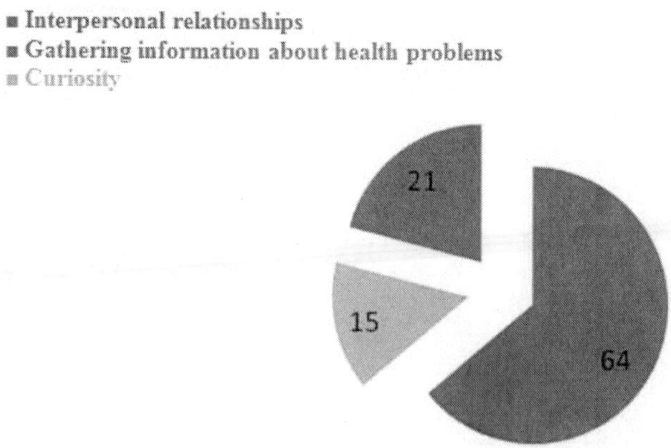

Most of the respondents choose to view *Grey's Anatomy* on the internet (86%) because they have the possibility to view more than one episode every day.

Figure 9-4. Do you view *Grey's Anatomy* on TV or on the Internet?

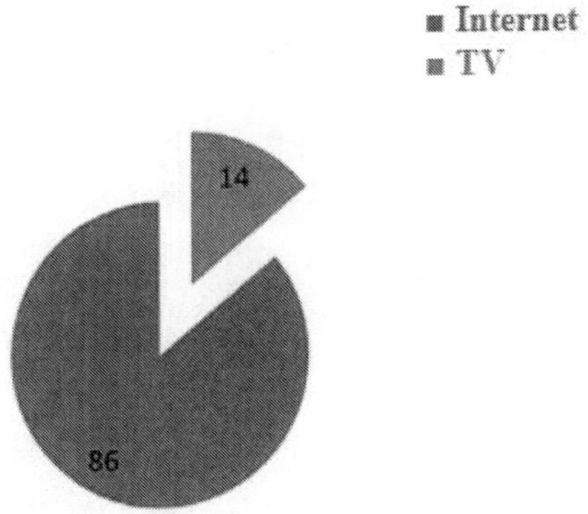

Although they consider themselves fans of *Grey's Anatomy*, the respondents cannot remember health care issues or cures presented by this show.

Figure 9-5. Do you remember a particular health care issue or health cure from *Grey's Anatomy*?

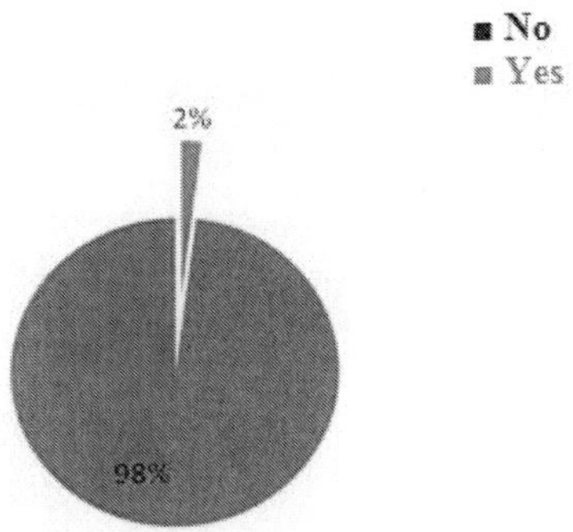

When it comes to the level of trust in the medical advice in *Grey's Anatomy*, the respondents declare that they do not trust the medical advice because they believe that the health information that it makes available to the public is incomplete, superficial or unreal.

Figure 9-6. Level of trust in the medical advice in *Grey's Anatomy*

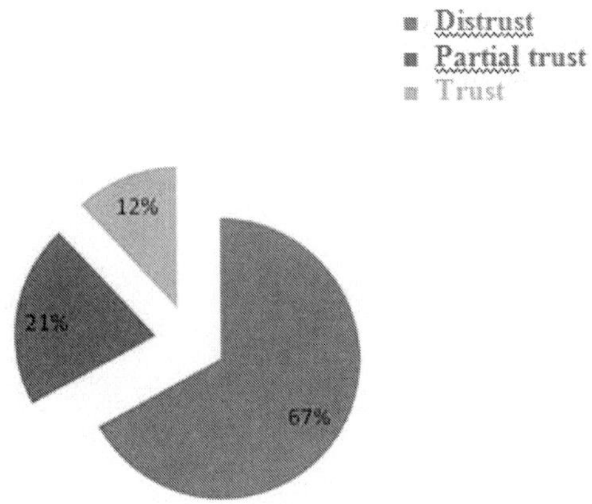

When it comes to influence on people's beliefs and behaviours, after viewing this show the respondents believe that to be healthy means having a good diet (95%) or being active and practising sports (87%). Feeling strong or being fit are also arguments for a person's good health.

Figure 9-7. What does it mean to be healthy?

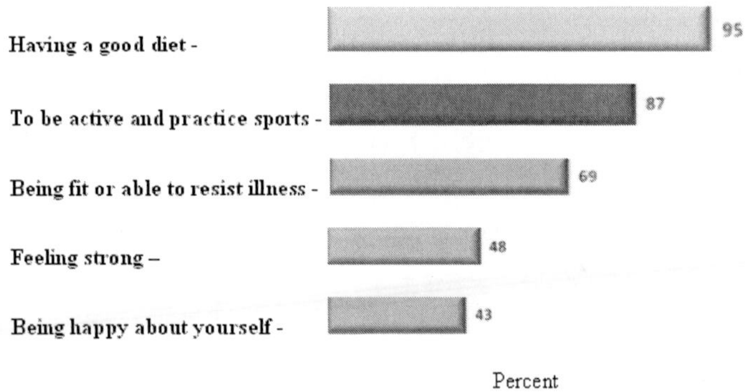

Almost all of the respondents consider that television plays a major part in shaping people's beliefs and understandings of health. It was clear from the discussions that the media is able to influence participants' views in less stereotypical ways. Therefore, for some respondents, the medical TV series were actually regarded as a positive influence which raised awareness of some important health issues such as cancer diet, exercise, or eating healthy food. This opinion comes in opposition to the scholars who state that the media and especially new media technologies have encouraged an increase in sedentary activities and unhealthy eating (Smith, Green, 2005).

Conclusions

Kellner argues that 'in modern industrial society a "media culture" has emerged which helps create the fabric of everyday life, shaping political views and social behavior, and providing the materials out of which people forget their very identities' (1995, p. 1). Kellner also points out that the products of cultural industries have become resources for the formation of personality models and therefore contribute to the construction and creation of cultural identities: 'radio, television, film, and other goods of the culture industries provide the models of what it means to be male or female, powerful or powerless. Media culture also provides the resources out of which many people build their sense of class, race and ethnicity, of nationality, of sexuality, of "us" and "them". Media culture shapes the prevalent view of the world and the deepest values: it defines what is measured good or bad, positive or negative, moral or evil. Media stories and images give the symbols, myths, and materials which help shape a common culture for the majority of individuals in many parts of the world today' (1995, p. 1). Therefore, the media is extremely powerful in promoting and creating role models for contemporary people. The media is extremely powerful in promoting health beliefs and in creating role models for contemporary people. It is clear that the media can play a significant role in shaping people's understandings of health, as well as in shaping people's health beliefs or food practices, especially through television programmes dedicated to health or through advertising. As Lang (1998) has also suggested, marketing and branding can change health beliefs or food culture over a relatively short period and most people are particularly influenced by these strategies. Medical TV series are an important source of health information for the public and they are regarded as a positive influence which has raised awareness of some important and controversial health issues.

References

Blaxter, M. & Paterson, E. (1982). *Mothers and Daughters: A Three-generational Study of Health Attitudes and Behaviour*, London: Heinemann.

Brashers, D. E. & Babrow, A. S. (1996). Theorizing health communication. *Communication Studies*, 47, 243-251.

Calnan, M. (1987). *Health and Illness: The Lay Perspective*, London: Tavistock Publications.

Curtis, S. (2004). *Health and Inequality: Geographical Perspectives*, London: Sage.

Curtis, S. & Taket, A. (1996). *Health and Societies: Changing Perspectives*, London: Arnold.

Eldridge, J. & Murcott, A. (2000). 'Adolescents' dietary habits and attitudes: unpacking the 'problem of (parental) influence'. *Health*, 4 (1), 25-49.

Jacobs, J. (2003). *Body Trauma TV*, London: British Film Institute.

Jones, L. (1994). *The Social Context of Health and Health Work*, Basingstoke: Palgrave.

—. (2000). What is health?. In Katz, J., Peberdy, A., Douglas, J. (Eds.), *Promoting Health: Knowledge and Practice*, Basingstoke: Palgrave.

Lang, T. (1998). The new globalization, food and health: is public health receiving its due emphasis? *Journal of Epidemiology and Community Health*, 52 (9), 538-39.

Lloyd, C. & Shakespeare, P. (2000). The Rise and Dominance of Biomedicine. In Team, K. C. (Ed.), *K203 Block 1: Visions and Values in Health Work,* Milton Keynes: The Open University.

Rideout, V. (2008). *Television as a Health Educator: A Case Study of Grey's Anatomy, A Kraiser Family Foundation Report*, California: Henry J. Kaiser Family Foundation.

Smith, A. & Green, K. (2005). The Place of Sport and Physical Activity in Young People's Lives and its Implications for Health: Some Sociological Comments. *Journal of Youth Studies*, 8 (2), 241-53.

Thomas, R. K. (2005). *Health Communication*, New York: Springer.

Williams, R. (1983). Concepts of Health: An Analysis of Lay Logic. *Sociology*, 17 (2), 185-205.

CHAPTER TEN

A POSTMODERNIST READING OF GREEK REALITY TV

ANASTASIA VENETI,
ACHILLEAS KARADIMITRIOU
AND STAMATIS POULAKIDAKOS

Cinema and TV are the dominant forms of art today, because it is in these visual media that the conditions of postmodern society are most thoroughly expressed
(Roberts, 2000: 45)

Digitalization, TV deregulation and an emerging new aesthetic have engendered new formats of television programming; among the most prominent are some new and more sophisticated reality shows. The current chapter constitutes a postmodernist reading of reality programmes by examining the case of Greek reality TV. Drawing on theories of postmodernity from Debord to Kellner, this study highlights those postmodern characteristics that have stamped indelibly this celebrated TV genre. What ensues from our examination is a typology of the Greek reality TV shows based on their morphological and thematic characteristics.

Reality formats derive from universal conditions such as family life, career seeking, love, and daily anxieties, featuring ordinary people as participants. As pointed out by several researchers of the field (Hadley, 2012; Holmes & Jermyn, 2004; Hill, 2005, p. 31) there are millions of viewers who tune in each week to watch real people in various private, public or competitive situations. Having adapted to contemporary trends and adopted more sophisticated aesthetics of 'liveness' [a term coined by Ferveit (1999, p. 794)], reality shows attract a growing audience. As Ferveit (1999, p. 794) argues in his illuminating study 'The focus on presenting the audiovisual evidence as much as "the story" is one of the features that distinguishes reality TV from earlier attempts to "catch the real". This focus also leads to an emphasis on the visible surface of the world rather than on deeper symbolic aspects.' The television industry maximizes each opportunity to increase its profit with new and low cost formats of reality shows. Therefore, we now witness new hybrid forms of reality shows and series that embrace drama (soap opera), documentary, film techniques, and game philosophy.

Features of the Postmodern Society

Postmodernity is associated with changes to institutions and conditions and with new social and political settlements and novelties globally. Technological innovations have brought massive changes in communication systems and the media landscape (Borgmann, 2000; Castells, 2000). Social theorists argue that today's world is organized by increasing globalization which reinforces the dominance of a world capitalist economic system (Harvey, 1990; Jameson, 1991; Kellner, 2003). For Kellner (2003, p. 11),

> … we are entering a new form of *technocapitalism* marked by a synthesis of capital and technology and the information and entertainment industries, all of which is producing an "infotainment society" and spectacle culture [….] Culture and technology are increasingly important constituent parts of global capitalism and everyday life in the contemporary world…

Following the breakdown of the Bretton Woods system and the decline of the nation-state, David Harvey (1990) identifies postmodernity with 'late capitalism'; a stage of capitalism characterized by a new mix of high-tech advances, the concentration of (speculative) financial capital and an increasing commodification and industrialization of ever more inclusive sectors of human life.

The capitalist supremacy over the nation-state has brought about radical changes in people's collective imaginary regarding the way it constitutes reality. We witness the mediatization of an important part of social experience and relations, proclaiming the shift from modern societies to media societies (Schmidt, 2001, p. 1). As early as 1967, in his prestigious study *Society of the Spectacle*, Guy Debord describes a media and consumer society organized around the production and consumption of images, commodities, and staged events. Postmodernity fuels additional forms of identity characterized by elements of instability and fragmentation. As Poster (1995, p. 24, p. 32) argues, we observe the emergence of a new individual identity that abandons its claims to rationality and autonomy. Although the postmodern political sphere is marked by new movements (feminist, environmentalist) and new approaches on issues such as civil rights and equality, the ideology of consumption has supplanted the opportunity for a more politicized confrontation of the world. Fredric Jameson (1991), one of the most prominent critics of postmodernism, argues that aesthetic production today has become integrated into commodity production generally. He described postmodernity as a condition driven primarily by the forces of multinational consumer capitalism. Jameson (1991, p. 17) also discerns a new kind of superficiality—'depthlessness'—that is evident in various stylistic features of postmodernism. This kind of superficiality can be spotted in the symbolic content that is produced as a consequence of the dominance of television and popular culture in postmodernity.

Since media deregulation, television content has abounded with popular entertainment programmes. Popular entertainment long had its roots in spectacle, which constitutes an ideal (profitable) commodity in the era of late capitalism. TV entertainment programmes constitute the most suitable terrain for promoting life-style, consumption, products, and identities. Various cultural forms that transcend national boundaries using novel technological advances manage to address the needs of an increasingly fragmented audience. The manufacture of celebrities and TV icons are used to further propagate and promote the products of consumer society. As Kellner (2003, p. 1) explains 'During the past decades, the culture industries have multiplied media spectacles in novel spaces and sites, and spectacle itself is becoming one of the organizing principles of the economy, polity, society, and everyday life.' Based on this observation along with Debord's assertion of the dynamic of media and consumer society, Kellner (2003, p. 3) suggests that media spectacle, by employing new media technologies and embodying society's basic values, has come to play an influential role in our everyday life—'seductive spectacles fascinate the denizens of the media and consumer society and involve them in the semiotics of a new world of entertainment, information, and consumption, which deeply influences thought and action'. Immersing themselves in this world, individuals are stupefied, and adopt an uncritical and a-political attitude. As it is best described by Debord (1967: Section 42), spectacle is a 'permanent opium war.'

In addition, postmodern society is identified with the waning of the real. Technological innovations and the economic conditions of the postmodern era, as described above, have given rise to a decentralized, media-dominated society, where the distinction between media and reality has collapsed. As Baudrillard (1994) explains in his theory of hyperreality, as postindustrial technology, particularly the mass media, becomes more integrated into our lives, the distinction between the real and the imitations of reality represented in these media is hard to achieve, while the meaning is systematically eroded. Baudrillard (1994) alleges that the postmodern world consists of simulations of reality, or hyperrealities, wherein signs refer not to an external reality but to other signs; having entirely lost their association with their referent. What we are left with are *simulations* of reality, that is to say copies of copies. Robins also points to a 'tendency to replace the world around us with an alternative place of simulation' (1996, p. 159). With reality giving way to hyperreality, our understanding of the world becomes less a product of subjective experience and more the result of the bombardment of images from popular culture.

Reality Shows on Greek Television: The life-cycle of a concept that never ends

On Greek television, reality shows represent a programming genre thriving in the frequencies of commercial television. The first programmes containing elements of the reality show genre can be traced back to the era of state monopoly on TV broadcasting (1966-1989). These were shows, retroactively grouped into the reality television world, that in essence predated the boom of this genre, which took place in the early 2000s. One of them was the well-known hidden camera show called *Candid Camera* (1972), based on small staged farces with simple, ordinary, unsuspecting passers-by as its protagonists. At the centre of attention were their spontaneous reactions and expressions that did not rely on acting preparation. Another harbinger of reality TV can be considered the talent-search show *Opportunity Knocks* (1977), a competition-based broadcast aiming at displaying young people with talent in the field of singing, dancing, acting, music composition, and song writing. When the state monopoly on broadcasting ended (1989), commercial television heralded a nearly reality format through broadcasts that featured unknowns disclosing their personal problems (*At last together*, 1992; *A New Beginning*, 1997; *Message Received*; 2000). That was just the prelude to a new television era (Table 10-1). These initiatory programmes continued to appear on TV screen—in new or revived forms—in later years as well (Table 10-1 and Table 10-2).

The merciless competition between commercial channels in order to attract the widest possible audience led the television industry to a programming policy wedded to market principles. Commercial television, operating in a multi-channel landscape, was frantically engaged in a struggle for the introduction of the most attractive television concepts flourishing abroad. Soon, switching on the uniformity would be identified as a key feature of the Greek television landscape, setting aside any promises of pluralism or variety in content. Reality shows constitute only one aspect of this uniformity, albeit a strong one, since they have now become an integral part of television programming.

What appeared on Greek TV over the 1990s as broadcasts of social reflection (featuring non-actors in unscripted situations exposing their problems) paved the way to the formal era of reality television, dating back to the early 2000s.[1] In Greece, this era started somewhat late, when abroad the concept of reality television had already enjoyed great success. It was a time when unprecedented TV shows (such as the groundbreaking one, *Big Brother*) became global franchises, spawning local versions in a wide variety of countries.

Reality shows on Greek television can be categorized into six types of broadcasts, based on morphological and thematic criteria:

First, there are the reality shows of confinement, where a number of strangers are challenged to stay imprisoned in a specific environment (usually a house) for an extended period of time. This kind of show pioneered many of the stylistic conventions that have since then become standard in reality television shows (e.g. interspersing of events on screen with after-the-fact confessionals on the part of participants, voting process, participants being eliminated per episode as well as the concept of immunity from elimination. All these elements are framed by unusual or dysfunctional cast members) (Table 10-3). Second, the reality shows of survival, where participants are asked to demonstrate that they can cope with difficult living conditions. Compared to the previous shows, these make more noticeable the idea of competition and elimination, since contestants battle against each other under extremely adverse conditions and are removed from the show until only one winner remains (Table 10-4). Third, there are the talent shows, where players cultivate their skills, usually while incarcerated in appropriate schools or colleges. These shows are typically composed of three phases of competition: initial open auditions, callbacks/participants selection, and live performance shows until the grand final (Table 10-5). Fourth, we come across the reality shows on the makeover theme. This make-over process may involve either the outward appearance of individuals or the setting of the material world surrounding them (such as homes or workplaces) (Table 10-6).

[1] According to Sorogas (2004: 27-28), reality shows of *Big Brother* type are the evolution of a number of past programmes, broadcast on TV worldwide: Latin-American telenovela or soap-operas, programmes of confessional character, domestic sit-coms, various versions of game/quiz shows, dramatized news-bulletins, talk-shows characterized by elements of popular journalism.

Fifth, the reality shows aiming to highlight relationships of emotional interaction among people in various environments. These are usually of competition character based on a variety of themes of special interest (Table 7). Sixth, there are the reality shows of advisory character with a view to bettering the lives of the people participating in them (Table 8).

Over the last 12 years, apart from the domestic versions of reality shows, they have also been broadcast on Greek television some original programmes of this genre, directly purchased from their countries of origin (e.g. *Extreme Makeover-Home Edition*, *Fear Factor*). Obviously importing a show is cheaper than producing it and therefore this turns out to be a clever complementary strategy, adopted by commercial channels' directors, in order to meet the public's lust for reality TV.

In Greece, reality television saw an explosion of popularity mainly in the first years of its life-cycle (2001-2005). The genre showed a decline over the last years of the 2000s; however this proved to be a temporary case. In effect, the television industry never let reality television fall into obscurity. More specifically, in the early 2010s although reality shows had reached a saturation point, television makers contrived new ways of re-launching and revitalizing the genre. The solution lay in the element of charity, which proved to be very effective in stimulating afresh the public's interest. The new format includes a song or dance contest, where professionals work frantically with show biz personalities (celebrities) in order to produce spectacular performances. Part of the revenues, derived from the public's voting process, are given to charities. Under the guise of charity combined with the glow of celebrities reality television is given a new lease of life. The Greek case demonstrates that even in a small television market reality shows can have a long life span as long as they renew their formats.

Table 10-1. Shows paving the way to the era of Reality Television

Programme Title	Year of Broadcasting	TV channel
Candid Camera	1972-1973	YENED
Candid Camera (revived form-remake)	November 1992–May 1993	ANT1
Hidden Camera (based on the same concept)	May–July 1993	ANT1
Candid Camera (revived form-remake)	1993–1994	STAR
Candid Camera (-//-)	2003	ALPHA

Opportunity Knocks (Na i efkeria) *Opportunity Knocks* (revived form-remake) *Opportunity Knocks* (-//-)	1977–1983 1989–1990 2001–2003	ERT MEGA MEGA
At last Together (Epitelus Mazi)	1992	ANT1
A New Beginning (Mia Nea Archi)	1997	STAR
Message Received	2000	ANT1

Table 10-2. Miscellaneous TV Shows borrowing elements from Reality TV. The main reality shows on Greek television (domestic versions of global franchises)

Programme Title	Year of Broadcasting	TV channel
The couple of the year	2002	ALTER
Worth Watching it (Aksizi na to dis) *Worth Living it (Aksizi na to zis)* *Talk (Mila)*	2006–2010 2010–2011 2011–2013	ANT1 STAR STAR
Final Decision	2006	ALTER
The Package (Pame Paketo)[2]	2006–2013	ALPHA
3 Wishes/ Life Like Dream	2006	ANT1
Anna Vissi: As long as I have voice (docu-drama)	2011	ALPHA
You are kidding (Plaka Kaneis)	2011	ANT1
Family of Calamity (Ikogenia tis Simforas)	2012	ANT1
All you are Hiding me (based on scripted reality concept)	2013	ALPHA

[2] *The Package*: The programme restores relationships for some reason disturbed over time or gives the opportunity to people with separated lives to meet again.

Table 10-3. Reality shows of confinement

Programme Title	Year of Broadcasting	TV channel
Big Brother 1 *Big Brother 2* *The Wall*[3] *Big Mother / Big Brother 3*[4] *Big Brother 4*	2001 2002 2003 2005 2010	ANT1 ANT1 ANT1 ANT1 ALPHA
Bar	2001–2002	MEGA
The Gym Show	2002	STAR
Party	2003–2004	MEGA
Mission	2003	ANT1

Table 10-4. Reality shows of survival

Programme Title	Year of Broadcasting	TV channel
Farm 1 *Farm 2*	2002 2004	MEGA MEGA
Survivor 1 *Survivor 2* *Survivor 3*	2003 2004 2006	MEGA MEGA MEGA
Survivor: Greece—Turkey	2006–2007	MEGA
Extra Large: 400 km on Foot	2007	MEGA
Patagonia: The Edge of the World	2010	MEGA

[3] *The Wall*: The house is divided in two parts, the luxurious one and the poor one. The two teams are constantly struggling about which one will pass its time in the luxury side.

[4] *Big Mother / Big Brother 3*: The show initially begins as *Big Mother*, in which nine housemates take part in the game with their mothers living together in the same house. Mothers cannot win the prize, but can stay with their children until they leave home. The concept proves unsuccessful and as a result the show changes its format halfway adopting the traditional *Big Brother* style.

Table 10-5. Talent shows

Programme Title	Year of Broadcasting	TV channel
Popstars	2002–2003	MEGA
Fame Story 1 *Fame Story 2* *Fame Story 2* *Fame Story 4*	2002–2003 2003–2004 2004–2005 2005–2006	ANT1 ANT1 ANT1 ANT1
Super Idol	2003–2004	MEGA
Dream Show 1 the Theatre *Dream Show 2 the Music* *Dream Show 3 the Music*	2004 2006 2007	ALPHA ALPHA ALPHA
So you think you can dance	2006–2008	MEGA
TV Stars Show *(TV Stars parousiaste)*	2007	ALPHA
Greece got Talent 1 *Greece got Talent 2* *Greece got Talent 3* *Greece got Talent 4*	2007 2009 2010 2012	ANT1 ANT1 ANT1 ANT1
X Factor	2008	ANT1
Next Top Model 1 *Next Top Model 2*	2009–2010 2010–2011	ANT1 ANT1
The Next Uri Geller	2009	ANT1
Top Chef	2010	ANT1
Dancing with the stars 1 *Dancing with the stars 2* *Dancing with the stars 3*	2010 2011 2012–2013	ANT1 ANT1 ANT1
Master Chef 1 *Junior Master Chef* *Master Chef 2*	2010 2011–2012 2012–2013	MEGA MEGA MEGA
Greek Idol 1 *Greek Idol 2*	2010 2011	ALPHA ALPHA
Just the two of us	2010–2011	MEGA
Dancing on Ice	2011	ANT1
Botrini's Project	2011	ALPHA
Your Face Sounds Familiar	2013	ANT1

Table 10-6. Reality shows of makeover theme

Programme Title	Year of Broadcasting	TV channel
Extreme Makeover	2008–2009	MEGA
Get with the Right Foot (Me to Deksi)	2010–2012	ANT1
Home from the Beginning	2010–2011	ALPHA

Table 10-7. Reality shows aiming to highlight relationships of emotional interaction

Programme Title	Year of Broadcasting	TV channel
A Bride for my Son	2009–2010	ALPHA
Love Bites	2009–2010	ANT1
Farmer Wants a Wife	2009–2011	ALPHA
Do not tell this to the Bride	2011	ANT1
The real housewives of Athens	2011	ANT1
Sunshine Girls	2012	ALPHA

Table 10-8. Reality shows of advisory character

Programme Title	Year of Broadcasting	TV channel
On your behalf (Gia logariasmo sas)	2009	ALPHA
Nightmare in the Kitchen	2009–2013	ALPHA
Hotel Inspector	2010	ALPHA
Super Nanny	2010	ALPHA
Fab 5	2011–2012	ANT1

Analysis based on the conceptual typologies of Greek reality TV

Reality television shows of confinement such as *Big Brother*, *The Wall*, *Bar*, *The Gym Show* represent a documentary-style programme, portraying the lives of unknowns in a sort of artificial coenobium. These different personalities are invited to develop relationships that turn out to be utilitarian and superficial, lacking any moral background. At least on Greek TV, quarrels and intrigues are an integral part of the fellowship developing among the incarcerated persons. This reflects Bauman's criticism of postmodernity, where he talks about fragmentary and casual relationships, driven by consumerism and without moral basis (Bauman, 1996).

In these shows, camera shooting and footage editing give the audience members the impression that they are secret observers, following people going about their daily activities. In this respect, the broadcasts are prime examples of Foucault's (1977) use of Bentham's Panopticon, where the power of surveillance is inexorable. According to participants, as time goes by, there comes a moment when the sense of camera and monitoring disappears, letting them be completely spontaneous. In this case, the whole setting reflects the concept of voyeurism, where the audience (the voyeur) is granted a spy view on a subject of his interest (group or individual) during intimate activities or actions considered to be of a private nature. More than this, in the Greek versions of the show, voyeurism takes place in its stricter definition. The audience (the voyeur) spy on participants without them being aware of it in places (such as toilets or storerooms) where even if they are not visible they are perceived thanks to the power of sound, which cannot be eliminated by any trick.

In order to attract a wider audience, the Greek version of Big Brother 2 incorporated in its flow improper scenes, giving rise to scathing criticism. In essence, the show in its second season crossed the boundaries of voyeurism and adopted elements of 'pornography' in the sense that it included transparent images or sounds that generated sexual reflections in the viewer. Too much openness (transparency) in everything verges on inappropriate results. What takes place is a kind of saturation of public and private spaces through obscene simulations, what Baudrillard (1985, pp. 130-131) calls 'the ecstasy of communication. All secrets, spaces and scenes abolished in a single dimension of information'. Because 'obscenity begins [...] when everything becomes immediately transparent, visible, exposed in the raw and inexorable light of information and communication' (Baudrillard, 1988, pp. 21-22).

In addition, reality shows of survival like *Patagonia* and *Survivor* owe part of their success to the unpredictability that is woven into this show genre. This is further enhanced by the fact that the programmes are filmed entirely in advance, which gives producers a leg up, enabling them to edit in overarching storylines, symbolism, and imagery. This manipulation of the construction of the show creates an aura of realness that engages viewers (Wright, 2004). In the words of Jean Baudrillard, 'it is rather a question of substituting signs of the real for the real itself' (Delisle, 2003, p. 44). Furthermore, the notion of competition inhabits these particular shows. The whole idea is about a game incorporated in a game where participants are devoted to constant challenges in order to reach the finals and ultimately to gain the cash prize. The emotion of competition is evident in the conversations of the contestants, their reflective monologues, and the formation of the voting alliances.

As part of the show's philosophy, the contestants must cater for their nutrition and shelter while sporadically they receive basic supplies. June Deery (2004), drawing on Marx's ideas on industrial fetishism and Baudrillard's notes in the *Consumer Society* (1998), describes the paradox of the contestants receiving these supplies as manna from heaven and not as the goods supplied by the production company. The abundance of goods is experienced as a miracle stripped of the historical and social condition. As Deery (2004, p. 14) argues, 'The contestants' situation is an exaggeration of our own as consumers.' These shows' multidemographic appeal, the buzz that is created around them along with the show's structure (Tribal Councils, casting votes etc) render them an ideal tout for product promotion in a simulated microsociety. In Deery's words (2004, p. 15):

> 'Producers encourage a totemistic consumption of associated by-products (T-shirts to towels) which helps define participation as consumption, a move reflective perhaps of the wider political sphere in which citizens are being replaced by consumers and in which interactive voting for reality show outcomes functions as an ersatz democracy.

Another aspect of reality shows is celebrity production, which is best exposed in talent-oriented reality shows such as *Fame Story* or *XFactor.* This new kind of ephemeral celebrity has changed the way we view and evaluate a genuine public figure. In effect, the real 'heroes' of the past have been replaced by the manufactured celebrities of reality shows, who may lack authentic talent. This replacement is clearly reflected in the arguments of Daniel Boorstin (1992, p. 13) when he describes a world stricken by a *Graphic Revolution*, where the omnipresence of images and pseudo-events has undermined the meaning of a famous public figure.

Talent shows with the aid of manufactured celebrities produce spectacle. The key feature of this spectacle is the emphasis on style over substance, on image over reality. The rising stars, sprouting from these shows, cultivate the idea that the dream of success is the main goal in life and everyone can treat it as an easily achievable task as long as he or she abides by the rules of spectacular showbiz. In this respect, the participants in these programmes are the 'mediation' of a number of false needs or false representations of life. As Debord (1994, p. 38) puts it, they represent ordinary people turning their spectacle into 'images of possible roles' for us to identify with so as to set aside real-life drawbacks of society (such as capitalist exploitation and class division). In that way, spectacle operates as an alienating scenario based on appearances-impressions, namely a false reality.

Reality shows of makeover format involve two subcategories. The first one aims at perfecting the outward appearance of people, while the second one deals with the improvement of their material world through home reconstruction. Both of them are based on economically unprivileged people who are given a style they cannot afford to sustain. The first subcategory displays contemporary standards of beauty acting as a vehicle for promoting consumer trends. In Toby Miller's (2008, p. 586) words, 'the makeover's power to fascinate is achieved through the ultimate consumer desire: self-invention via commodities.' In the second type of make-over shows, families from low walks of life are prominently seeking happiness in material goods. The key message is that material goods can trigger a new start in life. As a result, these broadcasts represent the triumph of commodities in the capitalist way of consumption. It is what Debord (1994, p. 44) calls the diffuse form of spectacle in the sense of 'pseudo-need imposed by the reign of modern consumerism [...] waves of enthusiasm for particular products, fuelled by the communications media, are propagated with lightning speed.'

Another popular category on Greek TV is that of reality shows of a socio-sentimental nature. Reality shows like *A Bride for My Son*, *Farmer Wants a Wife* and *Love Bites* focus on the search for the ideal partner. These shows allured both potential participants and viewers. Indicatively, 5,000 people applied for *Love Bites*, and 4,300 farmers applied for *Farmer Wants a Wife*, while the hostess of the show received over 2,000 letters from potential brides (Papanikolaou, 2013).

The focal point of this kind of real life programme is on the presentation of the emotional interaction between participants along with the exposure of their personal feelings. This is supported by the participants' self-reflections on camera, confessions, the footage from multiple cameras and the narration of the show's host/ess in some cases. This overexposure of the participants' sentimental condition serves to satisfy the viewers' needs - as consumers - for love, sex, companionship, secrets etc. High spectatorship ratings of these shows underpin the argument of Baudrillard's (1983) hyperreal condition and of the paradox of satisfying need for authenticity through simulation. As Firat (1992, pp. 79-80) argues 'It is most apparent in the preference of large segments of consumers for living and experiencing the simulation and the hype rather than the concrete reality.'

In addition, Annette Hill (2005) explains that people do not watch reality TV just for entertainment, but because they can also engage in critical viewing of the attitudes and behaviours of ordinary people in these programmes. As for particular target audiences, we support Deery's (2004, p. 5) claim that: 'It is tempting to understand this predilection in the light of the complexity and anonymity of postmodern life, with its notorious loss of community and coherent grand narratives.' These reality programmes capitalize on the interest in the emotions of other people which is predominant in contemporary Western culture. In the particular programmes, people reveal their emotions which, depending on the case, can be an amalgam of parallel reverence and ridicule among participants. In their study, Aslama & Pantti (2006, p. 168) examine 'reality television as an illustration of contemporary confessional culture in which the key attraction is the revelation of "true" emotions'; suggesting also that the use of monologue is the medium to approach authenticity.

Based on the above, we can assume that reality has become a highly marketable concept. In this context, there is another flourishing category of reality programmes of an advisory character. Taking advantages of people's personal and economic insecurities and problems of daily life, reality programmes such as *Nightmare in the Kitchen*, *On Your Behalf*, *Hotel Inspector*, *Fab 5*, and *Super Nanny* offer advice and suggestions for a better lifestyle, better child-rearing, more efficient financial management. According to the Greek daily newspaper *Ethnos*, 4,000 families applied to the programme *On Your Behalf* in order to get help and turn their fortunes around (Papanikolaou, 2013). The shows focus on people's difficulty in coping with their problems and as a consequence some episodes are highly emotional. As already explained, people's emotions and experiences become the epicentre of the media spectacle. Drawing on the ideas of Baudrillard, Kellner (2003, p. 20) claims that the postmodern media society exposes 'the most banal and intimate aspects of everyday life [...] or the melodramas of ordinary "real-life" drama participants.'

Concluding Remarks

Both the anatomy of the reality shows and the viewers' response confirm Jean Baudrillard's position that we are currently living in an era of simulation in which it is impossible to tell the difference between reality and simulation (1983; 1994). Reality TV achieves an access to reality through surveillance and voyeurism. It addresses primitive desires of the audience and their critical appetite as familiar situations can be easily criticized and received as entertainment—as 'the real' is displaced onto others.

In particular on Greek television, the greatest emphasis has been placed on reality talent shows, indicating that the audience thirsts for the idea of idols or rising stars seeking to realize their big dream. In general, television deregulation in Greece, taking place at the very end of 1989, contributed to the explosion of pop culture genres, familiarizing the audience with 'digestible' low-quality television content (Papathanassopoulos, 2005, pp. 303-305). Although excessive broadcast of these programmes led the audience quickly to saturation, producers' ability to generate innovative or unprecedented formats operated as a lifeline.

After all, it seems that reality television has no boundaries or has boundaries equivalent to the human mind's limitations. The secret lies in the charming effects of television realism on audience perspectives on the real world. As Fiske and Hartley (1978, pp. 161-162) aptly put it 'Television realism, then, following the pattern of language at large, "naturalizes" the way in which we apprehend the world out there [...] the world is, naturally and of itself, what the mind-originated conventions of realism say it is.'

References

Aslama, M. & Pantti, M. (2006). Talking alone. Reality TV, emotions and authenticity. *European Journal of Cultural Studies* Vol 9, no.2, 167–184.

Baudrillard, J. (1983). *Simulations.* Semiotext(e): NYU Pres, New York.

—. (1085). The Ecstasy of Communication. In H. Foster (Ed.) *Postmodern Culture* (pp. 126-134). Pluto Press: London.

—. (1988). *The Ecstasy of Communication.* Trans. B. Schutze & C. Schutze. Semiotext(e): NYU Press, New York.

—. (1994). *Simulacra and Simulation.* Trans. Sheila F. Glaser. University of Michigan Press: Ann Arbor.

—. (1998). *The consumer society* (C. Turner, Trans.). Sage: London. (Original work published 1970).

Bauman, Z. (1996). From pilgrim to tourist - or a short history of identity.In S. Hall and P. du Gay (Eds.) *Questions of Cultural Identity,* (pp. 18-36). London: Sage: London.

Boorstin, J. D. (1992). *The Image: A Guide to Pseudo-events in America*, Vintage: NY.

Borgmann, A. (2000). Society in the postmodern era. *The Washington Quarterly* 23, 187–200.

Castells, M. (2000). *The Rise of the Network Society. The Information Age: Economy, Society and Culture Vol. I.* Blackwell: Oxford, UK.

Debord, G. (1967). *Society of the Spectacle*. Black and Red: Detroit. [and 1994, New York: Zone Books].
Deery, J. (2004). Reality TV as Advertainment, Popular Communication. *The International Journal of Media and Culture* 2:1, 1–20.
Delisle, B. J. (2003). Surviving American Cultural Imperialism: *Survivor* and Traditions of Nineteenth-Century Colonial Fiction." *The Journal of American Culture* 26 (1), 42–55.
Fetveit, A. (1999). Reality TV in the digital era: a paradox in visual culture? *Media Culture Society,* 21, 787–804.
Firat, A. F. (1992). Postmodernism and the Marketing Organization. *Journal of Organization Change Management*, 5 no. 1, 79–83.
Fiske, J., Hartley, J. (1978). *Reading Television*. Methuen: London.
Foucault, M. (1977). *Discipline and punish: The birth of the prison.* Pantheon: NY.
Hadley, B. (2012). Self-Making as Public Spectacle: Bodies, Bodily Training and Reality TV. *Scope: An Online Journal of Film and Television Studies*, 24, 1-14.
Harvey, D. (1990). *The Condition of Postmodernity. An enquiry into the origins of cultural change*. Blackwell: Oxford.
Hill, A. (2005). *Reality TV: Audiences and Popular Factual Television*. Routledge: London.
Holmes, S., Jermyn, D. (eds.), (2004). *Understanding Reality Television.* Routledge: London.
Jameson, F. (1982). *Postmodernism and Consumer Society*. Retrieved from http://art.ucsc.edu/sites/default/files/Jameson_Postmodernism_and_Consumer_Society.pdf.
—. (1991). *Postmodernism, Or the Cultural Logic of Late Capitalism*. Duke University Press: Durham, NC.
Kellner, D. (2003). *Media Spectacle*. Routledge: London.
Miller, T. (2008). Afterword: The New World Makeover. *Continuum: Journal of Media and Cultural Studies*, 22 no.4, 585–590.
Papanikolaou, N. (2013). 51,000 Candidates in the Queue For a Reality Show. *Ethnos.gr* (July 10th 2013). Retrieved from http://www.ethnos.gr/article.asp?catid=22733&subid=2&pubid=9759001
Papathanassopoulos, S. (2005). *Television in the 21st Century*. Kastaniotis: Athens.
Poster, M. (1995). *The Second Media Age*. Polity Press: Cambridge, UK
Roberts, A. (2000). *Fredric Jameson.* Routledge Critical Thinkers. Routledge: London.
Robins, K. (1996). The Virtual Unconscious in Postphotography. In Timothy Druckrey (Ed.) *Electronic Culture: Technology and Visual Representation* (pp. 154-163). Aperture: New York.
Schmidt, J. S. (2001). Media societies: Fiction Machines. In Grant C.B., McLaughlin, D. Rodopi (Eds.) *Language–meaning–social construction. Interdisciplinary studies:* New York.
Sorogas, E. (2004). *The Phenomenon of "Reality"*. Kastaniotis: Athens.

Wright, C. J. (2004). Secret and Lies: *Survivor* and Frederic Jameson's *The Political Unconscious*. Thesis at the Faculty of the Graduate School of Arts and Sciences of Georgetown University. Retrieved from http://www8.georgetown.edu/cct/thesis/ChrisWright.pdf.

CHAPTER ELEVEN

SCOPING INTERACTIVITY: CONCEPTUALIZING THE POST-TELEVISION VIEWER

ANGIE CHIANG

The concept of the television audience has long been an elusive topic of intrigue in reception studies, entering a new phase of scholarship in the post-television era. In the network era, television content was broadcast en masse, resulting in regarding the viewership as a homogenous mass. During the cable era and facilitated by the advent of technologies such as the remote control and VCR, audiences began fragmenting into niche markets and viewer's choices became more relevant in the discussion of the television audience. As new media have come to define much of our contemporary post-television era, questions regarding the nature of the audience continue to persist, more particularly, those regarding how audiences interact with television.

Interaction is not a concept unique to post-television, as it has been much discussed (under varying terminology) consistently throughout not only television audience studies, but also in literary theory with notions of the 'active reader'. It has become a relevant topic in the post-television discourse namely due to the prevalence of new media in contemporary television. At its most basic level, 'interaction' refers to the reciprocal action or influence between sender and receiver. Traditional mass media or 'old media' (i.e. network era television) adhered primarily to a one-way communications model, where the audience would be unable to give timely or relevant feedback. 'Interaction' in the network era hinged primarily on the interpretative aspects of the audience and television text; similar to how a reader would understand a novel.

With 'new' media, defined as those emergent digital media forms or technologies that have been reprocessed or digitized by computers (Murphy, 2011, p. 11), 'interaction' focuses more on the user's (i.e. receiver) control as opposed to producer (i.e. sender) control in old media forms. Unlike interaction with old media, new media interactions can produce tangible markers that can be readily accessed and analysed online. Blog post reviews on television episodes, discussion forum threads or hashtagged social media messages are representative examples. That being said, notions of interaction in a post-television era do not exclude the interpretive conceptions that came before; in fact they continue to remain quite relevant in the current literature. 'Interactivity' is therefore best conceptualized not in monolithic terms, but as a concept that encompasses a spectrum of activities, from the interpretative to the physical actions of the viewers (Evans, 2008, p. 201).

Technology also plays an important role in the discussion of interaction in the post-television era, but often its usages have become so naturalized that they are overlooked. An inherent interaction exists on the simple level of choosing not only what to watch, but also when and how (i.e. on what platform) to watch it. These have become standardized practices for almost any television viewer, which means all but abandoning notions of a completely passive viewer. As interaction becomes more commonplace amongst the viewership, programme producers are directly calling on audience members to determine outcomes of shows. This is most acutely seen in reality television competition shows, which have flourished in the post-television age, allowing viewers to vote and to have a hand in eliminating contestants and choosing winners. Although competition shows offer an aspect of interaction that cannot be denied, the power of this action is still contained and controlled by the producers. There is an apparent opacity with shows of this nature, where the viewers place trust in the producers to deliver truthful outcomes. The call for interaction might be unquestionable, but the results of that interaction are still ambiguous. Furthermore, this interaction is bounded by the broadcast lifecycle of the programme; viewers are unable to interact on this level beyond a specified time limit and this interaction is all but irrelevant once the show has finished airing its season.

In contrast, complex prime time dramas exhibit a different kind of interaction, primarily found documented online on message boards, by way of fan fictions or on social media. *Lost* (2004-2010) one of the decade's most influential complex narratives, deliberately unfolds its narrative over non-linear fragments. Executive producers Damon Lindelof and Carlton Cuse stated that they intentionally developed the series in such a way as to encourage the participation of viewers. Cuse says:

The show became an excuse to develop a community…and the basis of it was that people were able to debate open-ended questions—questions that we insisted be open-ended, and that would get fans engaged in the show. (Cuse in Rose, 147)

Arguably, this empirical evidence of interactivity by way of online 'debates' exists for both dramas and reality shows, however those in relation to dramas develop a wider range of interaction due to the open-ended aspects embedded in the narratives. These dramas offer a larger range of sustained interactions and therefore serve as a more suitable case study in investigating the relevant literature's claim that the post-television audience is an increasingly interactive one.

The Internet is a large component in the discourse surrounding the post-television viewer, specifically, 'Web 2.0'. Fueled by social networking, blogging and microblogging, Web 2.0 is defined by the increased proliferation of user-generated content and the active calls for participation and sharing of content between consumers (Jenkins, 2006, p. 179). Web 2.0, or web commons as it is also referred to, does not speak to a change in the technology of the Internet, but a shift in the way we discuss its usage (Booth, 23). The shift saw the web no longer as information simply to be accessed, but a hub to create and disseminate *new* information. Ultimately Web 2.0 discourse seeks to develop a 'new, more empowered, more socially connected and more creative image of the consumer' (Jenkins, 2007, p. 358). Web 2.0 factors into the re-definition of the viewer in the post-television era as more of a user, who no longer just views the television content on screen but is engaging with it in tandem on the Internet.

A chief proponent in the discussion of the interactive post-television viewer is Henry Jenkins. Building from his seminal work on fandom, *Textual Poachers* (1992) an ethnographic study of media fan communities, Jenkins argues that the post-television viewer is essentially a fan. He notes that although terms such as 'fan', 'fandom', or 'fan culture', are not used in many of the discussions of media consumers in the digital age, these models 'rest on the same social behaviours and emotional commitments that fan scholars have been researching over the past several decades' (Jenkins, 2007, p. 359). For Jenkins, a 'fan' is a reader who 'appropriates popular texts and rereads them in a fashion that serves different interests; [fans are] spectators who transform the experience of watching television into a rich and complex participatory culture' (Jenkins, 2013, p. 23). The difference between the post-television fan and the one described in *Textual Poachers,* is that the latter holds a 'cult' status, acting on the fringes of mainstream practices and the former engages in some form of fan activity via the Internet (Jenkins, 2006, p. 142). New fandoms are emerging rapidly online in the form of websites, blog posts, social media interactions and discussion lists.

A key point in Jenkins's vision is the distinction between 'interactivity' and 'participation'. The former refers to the ways that new technologies have been designed to be more responsive to user feedback (Jenkins, 2006, p. 137). 'Interaction' in this sense, can include something as simple as changing the channel with a remote control, or playing a video game in an immersive environment. The key distinguisher is that the producer/ designer of the 'interactive environment' pre-structures the degree of interaction the user can partake in (Jenkins, 2006, p. 137). 'Participation' on the other hand, is much more open-ended, and is 'shaped by cultural and social protocols' where the control resides primarily in the hands of the consumers as opposed to the media producers (Jenkins, 2006, p. 137). He gives the illustrative example of conversations in a movie theatre, which are dictated by the tolerance of the surrounding audience members. The amount and volume level of talking during a movie in a theatre could vary depending on the social and cultural norms in which it is taking place, which conversely is *not* a product of cinema itself. With regard to the television viewer, Jenkins's focus is primarily on the participatory culture in which the viewer resides because of the interactive opportunities that surrounds him.

Integral to Jenkins's participatory culture is how convergence has enabled *collective* modes of reception in contrast to individualistic action. What was once considered a private activity, watching television, has become much more public, where online forums offer an opportunity for participants to share their knowledge and opinions (Jenkins, 2006, p. 26). Building on Pierre Levy's notion of collective intelligence, users pool their insights and information together to broadly circulate messages and opinions developing a wider network/community (Jenkins, 2006, p. 256). A collective intelligence develops through the flow of consumption, dissemination and discussion. We can see collective intelligence mobilized in projects like fan-based wikis, where users can instantly edit, add, or delete content from a large database related to a particular television show.

The increasing serialization in post-television narrative programming has mobilized Jenkins's notion of collective intelligence. As narratives have become increasingly more complex, they have moved away from episodic structures that featured contained narratives. The post-television drama instead favours cliffhangers, more intricate and sustained story arcs and inter-series mythologies. These dramas thus provide the opportunities to conjecture about what was not shown in a singular episode, propelling viewers to go online and discuss. While exploring fan motivations in *Textual Poachers*, Jenkins noted that the failure to satisfy in earlier arguably 'simpler' narratives is what resulted in fan activity, spurred by the desire 'to try and articulate for themselves and others [the] unrealized possibilities within the original works' (Jenkins, 2013, p. 24). The impetus for these fan articulations is easily transposed to the era of complex narratives. Since many gaps in comprehension are apparent when viewing a single episode, fans can expand their individual grasp on the narrative by pooling their knowledge with other viewers sharing theories, predictions, and the like.

The practices outlined in *Textual Poachers* still remain in the digital age; however other forms that have close ties to the collective intelligence join them. The new cultural objects are not necessarily attributed to a single author, but capitalize on a network of fans. A blog post, for example may have been composed by a single author, but is in a constant state of evolution because of the discussion it potentially spawns in the accompanying comments section. Comments often cause the author to edit or amend the original post or participate in the comments discussion at length. The lifespan of television blog posts specifically is somewhat limited as they are likely to be relevant only in and around the time of an episode's broadcast. But blog sites thrive on the consistent uploading of fresh content, which arguably continues a collective intelligence discussion as long as new content is produced.

Digital technologies have allowed for cult practices of fandom, such as re-watching and discussion, to be more readily available to post-television viewers where they can easily congregate online to part-take in conversation. The Internet has also caused practices of fandom to become increasingly immersed with the rhythms and temporalities of broadcasting. Some fans now go online to discuss new episodes immediately after the episode's transmission time or even micro-blog throughout broadcasts in order to demonstrate the timeliness and responsiveness of their devotion (Hills, 2002, p. 178). Such practices of viewing have not gone unnoticed by programme producers, many of whom are participating on social media outlets, engaging in direct communication with fans. Members of the creative teams are in direct contact with viewers, in addition to participating in the collective intelligence by producing official, studio-sanctioned paratexts. *Lost's* executive producers Lindelof and Cuse remained active on Twitter, continuing to respond to fan questions regarding the show long after its broadcasted run. During the show's airing, they hosted a weekly podcast discussing the show's overarching narrative themes after each broadcast to further cultivate discussion and debate amongst viewers. Glen Mazzara, showrunner for AMC's *The Walking Dead* (2010-2013) was also quite visible on social media during his tenure, tweeting out future plot developments, spoiler warnings and narrative teasers. He was also a regular guest on the show's follow-up show *Talking Dead*, a live talk show airing after *The Walking Dead*'s newest episode. The show invites viewers to call in and ask questions to creative staff and cast members, vote in online polls and stimulate Twitter discussions by using assigned Twitter hashtags.

Viewer/producer conversations were not unheard of during television's previous eras, however the difference in a post-television context shows these interactions as being far more profuse and immediate than the practices of the past. But as the boundaries between producer and consumer become increasing blurred in the post-television era, the polarizing notion that producers are profiting significantly off the efforts of fans has entered the discussion. The NBC drama *Heroes* (2006-2010), for example, launched a promotional campaign asking fans to submit their idea for who could be the next superhero. The winning idea would become a character on the show in a later season, where NBC would hold all creative rights to submitted ideas. Going beyond simply monitoring fan practices, this example illustrates the movement from complete stonewalling on part of the producers in protecting their copyrighted content to the active solicitation of and potentially profiteering off fan inspired works.

Although the co-opting of fan-produced cultural objects can teeter on the borders of exploitation, much, if not all of the content produced by fans is largely considered paratextual. There has yet to be an example of a 'canonical' core text that directly incorporates fan produced culture with its studio produced counterpart. Jenkins's participatory culture does offer a wealth of collective intelligence objects as found by way of wikis, fan-made merchandise, and lengthy discussion board topics, but there currently exists no concrete evidence that the consumer and producer roles have been completely amalgamated. Even so, Jenkins considers the fandom of the digital age as an 'experimental prototype or the testing ground for the way media and cultural industries are going to operate in the future' (Jenkins, 2007, p. 361).

Complex narrative scholar Jason Mittell acknowledges how Jenkins's forecasted collective intelligent participatory culture comprises only a minority of the post-television audience. The value of this smaller groups' efforts, however, are highly influential, and thus are more representative of the changing nature of the post-television viewer. Mittell assumes that the behaviours of the more active online fans are indicative of broader tendencies (Mittell, 2012) but his assumption is slightly problematic as he places much credence in the habits of this smaller subset as indicative of a significantly larger whole.

If we examine one of the best examples of collective intelligence, *Lostpedia*, a fan-created wiki site, the questionable notion of a representative minority comes under scrutiny. *Lostpedia* can claim some impressive statistics, with over 7,000 pages/articles and boasting over 25,000 registered users/assumed contributors. However, when we consider *Lost*'s viewership fluctuating between 10-17 *million* viewers, *Lostpedia* shows us that at best, less than 5% of the viewing audience is actively contributing to this fan paratext. The page-view statistics (i.e. the users of *Lostpedia* are not necessarily the contributors as well) are likely to be much more useful in determining a more representative idea of what *Lost*'s viewership is doing in a post-television context. Since its inception in 2005, *Lostpedia* reports over 150 million page views, although this is a problematic number in that the quality of viewers is unknown (i.e. are these unique page views or the same returning viewers? Does the site have a high bounce rate? etc.) still represents a closer percentage of *Lost*'s weekly viewership. We can safely assume that the users visiting the site and logging in page views are in some way familiar with the show and looking to enhance their experience of the narrative beyond the broadcasted episode. Since the content of *Lostpedia* is so specifically tied to the show *Lost,* visitors are most likely to be viewers of the show rather than random surfers.

Mittell and Jenkins are not incorrect in their assumption that the highly participatory group of fans is indeed a fairly influential group in the post-television era, but I do not believe that the focus on this particular group's habits can be generalizable to the post-television audience as a whole. The lack of a stringent minority/majority dichotomy in the digital age comprises Jenkins's defence in his focus on the highly participatory group of fans. He does not maintain that his definition is a generalizable to the post-television audience, but he neglects to address the large number of viewers who operate outside his definition. This concept of the participatory viewer does not take into account other *kinds* of viewers in the post-television era, who represent a dominant percentage of viewership based on the statistical data. After examining the numbers, perhaps we may conclude that Jenkins is hasty in proclaiming that the mainstream no longer exists. It would be less problematic if we considered Jenkins's highly participatory viewers to be fans who continue to be on the fringes, perhaps increasing in numbers, but still accounting for a small minority of the larger viewing public. More in-depth study is needed on the habits of the larger viewership that is lacking in the current literature.

If we consider those viewers who do not produce a tangible object as a result of their fandom (i.e. blog/discussion post, fan fiction, artwork, etc.) there is a tendency to consider them as 'passive consumers'. Passivity implies less effort, and Jenkins's insistence that the post-television viewer is invariably a fan, doesn't quite hold up to the reality. We should not lose sight of the more mundane, internalized articulation with media that characterizes a great deal of media consumption (Bird, 504), such as the practices surrounding the simple selection of content as mentioned previously.

Continuing with the example of *Lostpedia*, the number of page views would indicate that the vast number of site users are can be considered 'lurkers'. Also applicable to readers who view message board content and never comment, lurkers are essentially information seekers who could easily be considered active viewers but do not produce any empirical evidence of it online. The act of seeking out additional information in addition to being familiar with a complex narrative such as *Lost* should classify them as such (Costello & Moore, 2007, p. 134). Yet, these viewers hold a minor presence in the current literature. Unlike the fan that Jenkins's describes, these viewers are content with the consumption of their chosen series, happy to read about others' opinions or to seek out background information or spoilers (plot details that will be revealed in future episodes). Lurking on the net produces a more enlightening, discerning, insightful consumer who develops a heightened viewing experience over the viewer who simply views the text. Even if the number of lurkers increases, ultimately it yields very little change in the external products or commercial culture. But Jenkins's considers this viewer as simply 'on the way' to exhibiting the kind of participatory fan behaviour that he describes, stating that it is 'a slippery slope from being an information gatherer/lurker to full on participant' (Jenkins, 2007, p. 361).

We must also consider that much of the interactivity documented online is inconsequential and we should 'resist the urge to lump all participatory activity as evidence of a revolutionary change' (Bird, 2011, p. 505). Perhaps what is problematic about Jenkins's view is his reluctance to present a spectrum of interactivity, as mentioned earlier. Because we may examine the fruits of Jenkins's participatory viewer online, it is much easier to identify and study this user. An understanding of the way a more 'passive' viewer interacts with a television text can likely only be gleaned from an ethnographic study. Although the ethnographic facet of scholarship is slower to catch up with generalizing theories, we need to take into account the vast range of online participation that is currently taking place. It is also important for us to consider that much of the fan interaction online is transient. The wikis, for example, could change the moment a given person, fan or not, chooses to alter the text. Blogs could pop up and subsequently disappear due to poor viewership or lack of interest on the part of the author. Incorporating various levels of fan interactions over time might provide a more representative conception. The interpretive activities of a viewer who simply watches a program, should be equally considered. Although, arguably this interpretive interactivity could apply to any text, the opportunities found in complex narratives are far more abundant than in a simpler narrative.

For example, if we examine the narrative presented on a reality television show like *Survivor*, a viewer could be interacting with the text on the level of wondering who will be voted out by the end of the episode, or even perhaps who will win the game based on the strategies employed. The questions posed by the audience are likely to be answered during the course of the episode, or at the very latest, by the season finale. If we contrast this with a complex narrative such as *Lost*, in the pilot episode alone, the viewer is left to ponder much more sophisticated questions such as 'What caused the plane crash?', 'Will the survivors get off the island?', 'Who is already on the island and what is their agenda?', 'What is the smoke monster?', and 'Why is there a polar bear on a tropical island?'. The answers to these questions are slowly rolled over the course of at least three seasons, some even stretching to the series finale (airing over six years later), or they may be denied an answer outright. Additionally, the significant gaps in time between broadcasted episodes require sustained comprehension as episodes break on a weekly, sometimes monthly basis, and especially when a show goes on hiatus during the summer months. Due to their resistance to adhere to closure-based, episodic structures, complex dramas mirror the kind of psychological interaction that we encounter in classic literature where readers are called upon to fill in any conceptual blanks in prose. One could argue that going online and reading interpretations of events constitutes one manifestation of bridging the gaps between texts. The lurkers who visit participatory websites but do not participate could be attempting to organize hypotheses that might be taking place before the text continues after a break or hiatus.

Interactivity in the post-television era should also encompass the ways in which the requisite technologies of television have affected our viewership. Often the simple act of perusing a television schedule, clicking a remote control or setting up a DVR aren't considered in the spectrum of interaction. Sheila Murphy (2011) emphasizes the importance of including the technologies of television in the discourse of interactivity. She describes the reciprocal nature of the technologies as mediators allowing us to personalize how and when we interact with the content. If, for example, a viewer wanted to dissect a particular scene by rewinding it and playing it in slow motion, the remote control would have significant impact in actualizing the interaction. The choice to 'binge-watch' an entire series over DVD could also produce interesting data on the nature of spectatorship. Although the use of such technologies isn't necessarily unique to the post-television viewer, the naturalization of the *use* of such devices is, and should be considered in the spectrum of interactivity activities of post-television.

The various theories that constitute television audience studies tell us that striving towards a unified theory of reception is a herculean task. With the fragmentation of the audience in a digital age, finding a generalized conception of the post-television viewer is proven even more difficult. It is far more useful to *incorporate* the characterization of the participatory viewer as Jenkins provides it, developing a *spectrum* of interaction that encompasses the various media that comprise contemporary television and the various forms of interactions that fall within. User interactivity has proven to encompass a range of different uses and agents, and it is extremely relevant to develop a more nuanced model for understanding its cultural complexity than the more generalized conceptions we have at present (Van Dijck, 2009, p. 46). Post-television reception studies should also consider the audience in two ways: not only as the collective that Jenkins understands, but in the tradition of Morley (1992) and Ang (1982), as a collection of individuals. How do the consumption patterns of viewers who continue to watch television 'in the background' while preparing a meal in the kitchen factor into our conception of the participatory viewer? What of the viewer who updates their social media status offering no additional insight on a programme aside from their approval of it? What about the discussions that arise from anti-fans of programs? It seems that to find an inclusive conception of the post-television viewers would be to acknowledge them as capable of inhabiting multiple and mobile identities that fluctuate from situation to situation (Ang, 1991, p. 162). Perhaps breaking down the audience into consumption patterns of specific television texts or niche groups is the only way to discover a representative picture.

Paradoxically, when looking at the individual viewer's interaction with a text we must also consider how significant cultural categories such as gender, race, and class, which are abundantly diverse when examining audiences in urban centres, affect this interaction. The heterogeneity of the television audience has always been a prohibitory in finding a unified definition over the years. As such, the conception of the post-television viewer must be a multilayered conception which accounts for both individual practices and how those are conceptualized in broader strokes. Conceptualizing the post-television viewer does not completely abandon the content of television text but requires that we study it in tandem with its usage in everyday life. Media use is just as important as the content being consumed. The current claim that the post-television viewer is inherently a more active participant is one worth investigating but we must consider the multi-faceted environment that created that viewer.

There is no doubt that technological and economic convergence has drastically changed television over the past 20 years. The post-television viewer is situated squarely in a digital context, but changes are not only evident in the nature of texts available but have also caused an entirely new set of naturalized media consumption habits to be adopted. Invariably, in order to establish an inclusive definition of the interactive post-television viewer we must examine the surrounding environment which both the viewer and text inhabit. Jenkins's notion of the participatory viewer is a large component of the post-television viewership; however, more pointed ethnographic study is required to more adequately develop the concept of 'audience' in the digital age.

References

Ang, I. (1982). *Watching Dallas: Soap opera and the melodramatic imagination.* New York: Methuen.

—. (1991). *Desperately seeking the audience.* New York: Routledge.

Bird, E. S. (2011). Are we all produsers now? Convergence and media audience practices. *Cultural Studies.* 25(4-5), 502–516.

Booth, P. (2010). *Digital fandom: new media studies.* New York: Peter Lang Publishing Inc.

Costello, V. & Moore, B. (2007). Cultural outlaws: An examination of audience activity and online television fandom. *Television & new media.* 8(2), 124–143.

Evans, E.J. (2008). Character, audience agency and transmedia drama. *Media, culture, society.* 30(2), 197–213.

Hills, M. (2002). *Fan cultures.* New York: Routledge.

Jenkins, H. (2006). *Convergence culture: Where old and new media collide.* New York: New York University Press.

Jenkins, H. (2007). Afterword: The future of fandom. In Gray, J., Sandvoss, C. & Harrington C.L (Eds.) *Fandom: Identities and communities in a mediated world.* (pp. 357–364), NYU Press, New York.

Jenkins, H. (2013). *Textual poachers: Television fans and participatory culture (20th Anniversary Ed.)* New York: Routledge. (Original work published 1992).

Mittell, J. (2012). *Complex TV: The Poetics of Contemporary Television Storytelling.* (In progress). Retrieved from http://mediacommons.futureofthebook.org/mcpress/complextelevision/

Morley, D. (1992). *Television, audiences and cultural studies.* New York: Routledge

Murphy, S.C. (2011). *How television invented new media.* London: Rutgers University Press.

O'Reilly, T. (2005, September 30). "What is Web 2.0? Design Patterns and Business Models for the Next generation of software." Retrieved from http://oreilly.com/web2/archive/what-is-web-20.html

Rose, F. (2011). *The art of immersion: How the digital generation is remaking Hollywood, Madison Avenue, and the way we tell stories.* New York, NY: W.W Norton & Company.

Van Dijck, J. (2009). Users like you? Theorizing agency in user-generated content. *Media, culture & society.* 31 (1), 41–58.

CHAPTER TWELVE

'YOU HAVE TEN FINGERS, I HAVE TEN FINGERS, LET'S BE FRIENDS': THE POST-MILLENNIAL FAMILY IN BRYAN FULLER'S *DEAD LIKE ME* AND *PUSHING DAISIES*

VERA CUNTZ-LENG

Introduction

As a social construct, family has been the subject of constant changes and reinventions, mirroring larger developments within society. Likewise, the medial representations of the family, especially on television, have been both the discussion of larger societal developments and the blueprint for emerging trends in our perception of family. In her study on the representation of the American family in prime-time television, Taylor (1989) argues that television is not a photocopy of the contemporary societal climate but 'speaks to our collective worries and to our yearning to improve, redeem, or repair our individual or collective lives, to complete what is incomplete, as well as to our desire to know what is going on out there in that elusive "reality." Television comments upon and orders, rather than reflects, experience, highlighting public concerns and cultural shifts' (p. 3).

Bryan Fuller's *Dead Like Me* and *Pushing Daisies* are particularly suitable subjects for the analysis of the post-millennial television family because of their supernatural theme, their high artificiality, and their high degree of abstraction. Both programnes overcome realism to challenge the conventions that society and television have established regarding the understanding of family. Furthermore, several other post-millennial television programmes also offer alternatives to the nuclear family model, and can complement the argumentation—like the popular sitcom *How I*

Met Your Mother or the supernatural drama series *Heroes*, among others. Within these shows, various family models are portrayed; and the dysfunctionalities of the biological nuclear family repeatedly force the main characters to find alternatives to this concept.

These new types of surrogate families can be put into context with the so-called 'chosen families' from the 1990s, as portrayed in *Buffy the Vampire Slayer* or *Friends*, to outline an important shift towards 'expedient alliances' as main family replacement in the post-millennial television series. The current surrogate families are defined by strong contradictions: the distrust in conventional families expressed through frequent ironic deconstruction and a nostalgic longing for an idealized version of precisely these traditional family relationships. At the same time, the main characters have become too dysfunctional themselves to ever achieve their ambition. Therefore, recalling Taylor, it is open to question whether the post-millennial television series still seeks for improvement, reparation, and redemption for the needs and lacks that the individual experiences in Western societies.

On Family

Family formations have always been the subject of changes and reformulations. Furthermore, their specific concept is highly dependent on the cultural background in question. Speaking generally, the term 'family' refers to a group of people that either share biological kinship, are related by marriage, or are united through co-residence. The dominant idea of family within Western societies is the conjugal or nuclear family, consisting of the male husband as principal earner, his female (stay-at-home) wife, and their biological children, although in reality, the nuclear family is not the dominant lifestyle. Especially the constant increase of single-person households since the 1970s (Euromonitor International, 2012) invalidates any assumptions that the nuclear family could operate as a lifestyle prevalent enough to be identified as dominant. At present, no exclusive concept of the family but a vast diversity in current family structures is detectable. Matrilocal families (mother and her children), patrilocal families (father and his children), consanguineal (extended) families, stepfamilies, adoptive families, families with same-sex parents, local residential communities, so-called empty nest families, and single-person households (Baca Zinn & Eitzen, 1987; Coontz, 2000; Coontz, 2005) are a few examples of today's diversity in family arrangements. Stacey (1996) introduces the term 'postmodern family' to give consideration to this great variability in family forms.

Notwithstanding, a societal nostalgic tendency to idealize the nuclear family is prevalent. Baca Zinn and Eitzen (1987) assert that the traditional family is generally viewed as a fulfilling and protective haven from the outside world, which compensates for the lack of other missing social connections. The family is 'a place of intimacy, love and trust where individuals may escape the competition of dehumanizing forces in modern society' (p. 3). According to Baca Zinn and Eitzen, popular wisdom ignores the fact that the family structures of the past were not superior to those of the present day. They claim that 'there is no golden age of the family gleaming at us in the far back historical past' (p. 8). Families were not more stable or happier. Singledom (spinster/eternal bachelor), infidelity, desertion of a spouse, illegitimate children, or homosexuality were issues then as well.

Particularly in the 1950s and 1960s but—after a more freethinking disruption in the 1970s—again in the 1980s, the representation of the family on television generally contravenes this observation. This underlines the significance of television's function as preserver of a specific salutary vision of the family in the collective memory. This can be confirmed by recalling Taylor's (1989) argument that television is not a photocopy of the contemporary societal climate but instead a spokesman addressing "our collective worries and […] yearning to improve, redeem, or repair our individual or collective lives, to complete what is incomplete" (p. 3). Popular family programmes of the 1980s like *The Cosby Show*, *Roseanne*, *Growing Pains*, or *Family Ties* explicitly ignored diversity in families to maintain the illusion of the happy, homogeneous nuclear family, 'burying their heads in the nostalgic sands of "traditional values" that never were' (p. 167). But meanwhile, other programmes like *Who's the Boss?*, *Full House*, *Blossom*, or *Married... with Children* tentatively started to deal with the postmodern family, paving the way for to-the-core postmodern programmes such as *Dead Like Me* and *Pushing Daisies*.

Dysfunctionalities of the Traditional Family

Mortality, afterlife, and death are the major tropes of *Dead Like Me* and *Pushing Daisies*. The protagonist of *Dead Like Me* is the college dropout Georgia 'George' Lass who lives with her parents and her younger sister Reggie. In the first episode, George is killed by a toilet seat that drops from the MIR space station while she takes her first lunch break from a new job. Instead of being dead, she is henceforth obliged to collect the souls of people who are about to die. With the cross Rube as her new supervisor, George is assigned to a group of grim reapers dealing with accidents, suicides, and murders: the mentally instable drug addict Mason, the rough police officer Roxy, and the fun-loving Betty (replaced by the self-centred former actress Daisy in 'My Room' (1.06)). They meet daily at the same restaurant Der Waffle Haus. Although dead George now has an occupation as reaper, she still needs to make a living in a day job. This brings her back to the very same temporary employment agency she had worked for the day she died. Being dead, George looks different to her former colleagues and family members. Rube, however, forbids her to interfere with the issues of the living and warns her not to get too close to her old family.

Similar to Rube, the mysterious pie-maker Ned of *Pushing Daisies* forbids his childhood sweetheart Charlotte 'Chuck' Charles to get too close to her aunts Lily and Vivian who think she has passed away. Technically speaking, they are right: Chuck was murdered on board a cruise ship and ended up in a funeral home. In the attempt to solve her murder mystery, private investigator Emerson Cod and his supernatural assistant Ned come to the funeral home. Ned, who has the special ability to temporarily wake the dead with the touch of his finger, recognizes Chuck as his painfully missed childhood love. Once he has touched her, Ned refuses to touch her again, which would kill her ultimately. In consequence, Chuck is alive, and a minor character drops dead as a substitute, Chuck and Ned can never touch, and she must not see her beloved aunts ever again or tell anyone about her existence, thus guaranteeing the continuance of the secrecy of Ned's ability. After her resurrection, Chuck moves into Ned's apartment, which is located over his restaurant The Pie Hole and henceforth—together with Cod and Ned's co-worker Olive—she becomes a valuable part of the crime solving team.

The constant usage of George's voice-over in *Dead Like Me* is an effective strategy to establish her status as first-person-narrator and focalizer (Branigan, 2007) for the audience. In *Pushing Daisies*, the identification of the main focalizer is more confusing since an omniscient narrator is superordinated to the story, following the narrative tradition of fairy-tales. But since George and Chuck have a lot in common—they both die and live again without having a hand in the matter, they share male nicknames to signify their inability to reproduce, both characters undergo the most significant transition in their respective series—it is justified to declare Chuck as the audience's main focalizer in *Pushing Daisies*, the so-called 'primary character-agent' (Branigan, 2007, p.73). This information is most useful to underline the importance of both characters, Chuck and George, for the audience experience of the respective programmes: when perceived through their lens, the dysfunctional representation of the traditional family becomes obvious, with both characters suffering from exclusion, loss, and incompletion, and carefully trying to establish new ties.

Like Chuck and George, the main characters of recent television shows usually come from neurotic and dysfunctional backgrounds. In fact, they are deeply troubled themselves, and in the process of a constant search for belonging. The singledom panic of the 1990s, on the one hand, which was omnipresent in programmes like *Sex and the City*, and the new consciousness of alternative family concepts during the last decade, on the other hand, paved the way for a variety of representations of the dysfunctionality of the traditional nuclear family on television.

The most widespread dysfunctionality is the absence of a parent. The Oedipal trauma of the absent father as experienced through the main character of a story has been a major trope throughout cultural history, which has recently been revived in various television series. A few examples are the fathers of the three main male characters of *The Big Bang Theory* Howard, Sheldon, and Leonard, who either disappeared in their early childhood or have so little influence that they never physically appear. Also J.D. in *Scrubs* ('My Cake' (4.06)), Marshall in *How I Met Your Mother* ('Bad News' (6.13)), and the whole Fisher family in *Six Feet Under* ('Pilot' (1.01)) are severely traumatized by the death of the father. *Lost* (Jack, Kate, Sawyer, Locke, Claire, Miles, Daniel) and *Heroes* (Peter, Nathan, Sylar, Micah, Mohinder, Matt, Hiro) are populated with characters who suffer from the loss or absence of the father. Moreover, the traumatization of the brothers Alan and Charlie through a disappeared Dad and an overpowering mother is the main reason for an endless list of failing romantic relationships and trite jokes in *Two and a Half Men*. In other examples, such as *Queer as Folk* or *True Blood*, the absence of the father is not a major issue but a subliminal part of the protagonist's biography, which influences their vision of and their trust in traditional family arrangements.

The ways in which the dysfunctionality of the nuclear family is depicted in *Dead Like Me* and *Pushing Daisies* is of particular interest because both programmes are defined by the addition of a bitter, dark irony that is not as eye-catching in the other examples mentioned above. Through several cutbacks, we learn that family life in *Dead Like Me* had been a burden for George since the birth of her sister. At the day of her death, George may be part of an intact American middle-class family but the menace of her exclusion hangs in the air. In the pilot (1.01), her mother Joy wants George to find a job and move out of the family household. For the upcoming job interview, Joy throws formal black clothes on her daughter's bed. Surprised, George remarks that these are funeral clothes. Joy—whose name is, regarding the sour nature of the character, an absurd prevarication in itself—replies dryly that there will 'be a funeral if you don't get your ass out of bed.' Ironically, these are the last words, Joy will ever address to her daughter before the fatal accident. And even more ironically, the Lass family haven will fall apart after George's death. After constant struggles in the first season and their inability to deal with their loss, they file for divorce in 'Send in the Clown' (2.01). In the later produced movie *Dead Like Me: Life After Death*, it is finally Joy who has to leave the house herself, recalling her past threat towards George from the first episode.

In consequence, when nostalgic longing over and over again pulls George back to her family home, it is the longing for something that does not exist anymore or maybe never existed, regardless of her being dead or alive. George desires the utopian idea of the happy nuclear family, not the family reality, underlined in particular by her growing sympathy for her formerly annoying sister, the need to talk to her mother, and the increasing discrepancy between George's envisioning of happy family life in several flashbacks and the bitter reality of her parents' separation. As mentor, Rube constantly explains that George has to overcome this desire for an invented past. Meanwhile, her parents and Reggie have to face the same obstacles, namely the collapse of the nuclear family formation. Shapiro (2011) and Moore (2006) have both referred to George's task of accepting death and her afterlife relationships, but in addition, this process is paralleled by the developments within George's biological family. The Lass family continues in their work of mourning and in their attempts to reorder their ways of living while it seems that George has found a workable alternative in the company of the other grim reapers. Therefore, the inability of George's biological family to cope with loss is mirrored in George's gradually evolving ability to move on, because she finds a new sense of belonging in the surrogate family of reapers (Gill, 2011).

The irony of *Pushing Daisies* is even more painful for the protagonist. When she is only eight years old, Chuck's father is accidently killed on her birthday by her friend Ned. In a hopeless attempt to keep alive his own mother, who has dropped dead from a heart attack, Ned touches her. When the crucial minute is over, Chuck's father dies instead of Ned's mother. On the same evening, Ned's mother kisses him goodnight and his touch kills her forever. It is a bitter irony that Ned and Chuck lose their respective parent the very same day because of Ned's ability to bring back the dead or rather kill ultimately. Even more ironically, both parents are killed by the ability that will give Chuck a second life over a decade later.

After this tragic day, Ned not only becomes a motherless child but he is abandoned by his own father and sent off to a boarding school. Meanwhile, Chuck moves in with her sociopathic aunts, who never leave their home. For both children, the dream of the nuclear family haven is over. Unbeknownst to Chuck and Vivian, aunt Lily is in fact Chuck's biological mother who constantly dismisses this relation and tries to keep it secret as it would endanger the continuance of the new, alternative family model that she, her sister, and Chuck represent. After the loss of Chuck, the dysfunctionality of this triangle becomes as evident as the problems of the Lass family in *Dead Like Me*.

As with George, Chuck's supernatural rebirth is combined with alienation from the concept of family she had formerly lived by. Separated from her aunts, Chuck finds an alternative group of peers in Ned, Cod, and Olive. Dauphin (2011) declares that the group identity of the 'Pie Holers' forms because of their isolationist personalities that separate the four characters from the rest of society. Chuck and Ned may have lost their biological parents, but Olive also had a troubled childhood ('Window Dressed to Kill' (2.11) and Cod is a counterpart to the absent father—an abandoned parent who suffers from the loss of his disappeared daughter. Moreover, it is revealed in 'Frescorts' (2.04) that Cod's own mother has fair skin while he is of African-American lineage. Though never debated publicly, this addresses on a subtextual level both an implied dysfunctionality in Emerson's biography and the general penetrability of family conventions and expectations that the series constantly challenges. Given their respective biographies, one could argue that all four characters are actually unified by their shared experience of the malfunction of conventional family structures. Since this encourages testing counter concepts, these individuals produce in the Pie Hole what Halberstam (2005) has described as 'queer counterpublics' (p. 6).

With an awareness of the trope of dark irony in the depiction of the nuclear family in post-millennial television series, it becomes more obvious that other examples operate in similar ways as *Pushing Daisies* and *Dead Like Me*. Isn't it ironic that Claire and Jack of *Lost* travel on the same plane with their deceased father in the cargo hold, not knowing that they are half-brother and sister? Isn't it ironic that Sylar of *Heroes* feels guilty for killing his mother in 'The Hard Part' (1.21) when it later turns out that they were not biologically related and he had been told lies about his real parents all his life? Isn't it tragic, ironic, and absurd at the same time that Barney in *How I Met Your Mother* first believes that show master Bob Barker is his father ('Showdown' (2.20)) and is later desperate to become a functional part within the family reunion in 'Cleaning House' (2.06) when his Afro-American brother finally finds his biological father? How ironic that Barney is totally disappointed when, four seasons later, in 'Legendaddy' (6.19), he finally meets his biological father, who is a mediocre driving instructor from suburbia.

Chosen Families and Expedient Alliances

Several studies (Battis, 2005; Burr & Jarvis, 2007; Busse, 2002; Sandell, 1998; Stevens, 2010) suggest that certain programmes of the 1990s like *Buffy the Vampire Slayer* or *Friends* give a chosen network consisting of friends preference over the traditional nuclear family. In their essay *Representations of Alternative Lifestyles in Buffy the Vampire Slayer*, Burr and Jarvis (2007) argue that *Buffy* 'explores both the advantages and the dangers of non-normative family forms, specifically the non-genetic or "chosen" family. [...] *Buffy* endorses a non-hierarchical, "democratic" vision of the family [...] with relationships between friends at its centre' (p. 263). Stevens (2010) believes that the 'fracturing of the traditional family unit results in the expansion of other relationships to compensate for its absence, manifesting in the adoption of friends as family" (p. 7).

Battis (2005) reads more recent programmes like *Queer as Folk* and *Sex and the City* in a similar way; but I would argue that this assumption falls too short, since many of the post-millennial television series have another parameter setting for the alternative families they portray. Neither the grim reapers of *Dead Like Me* nor the company of isolated individuals in *Pushing Daisies* are exactly friends. The same is true for the vampires, werewolves, shifters, and fairies in *True Blood*, who cannot choose their lineage and are sometimes forced to subordinate themselves under certain familial hierarchies. Moreover, the people with supernatural abilities in *Heroes*, the spaceship crew in *Firefly*, or the victims of the plane crash in *Lost* are bound together by coincidence rather than by choice. Even the nerdy colleagues of *The Big Bang Theory* are not essentially united by friendship but connected on a professional level as scientists and co-workers. This small distinction between the chosen families known in *Friends*, *Buffy the Vampire Slayer*, *Sex and the City*, and *Queer as Folk* and the 'expedient alliances' of later programmes is crucial, since it signifies a shift in the representation of alternative families in post-millennial television that emphasizes subject-matters of a postmodern way of life.

According to Sandell (1998), the characters of *Friends* have 'no control over their professional or romantic lives, the one thing these characters can control and change are their friendships, and hence the family community these friends become' (p. 148). This statement is not valid for *Pushing Daisies* or *Dead Like Me*. On the contrary, following a quite deterministic worldview, the characters are put together involuntarily, provided only with the freedom to choose what to do with their new relationships while their being together is set into stone. This can be seen as a counterpublical imitation of the biological family that is bound together by blood, not by free will. Although Dauphin's (2011) analysis of the formation of the group identity in *Pushing Daisies* might work in part for Ned, Emerson, and Olive (who are still also co-workers), being dead and forced to hide like Chuck as the audience's focalizer cannot be minimized as an 'isolationist quirk' (p. 29). Imprisoned by Ned in a golden cage, her affiliation with the 'Pie Holers' is as involuntary as George's assignment as grim reaper in *Dead Like Me*. Gill (2011) states that 'George is stuck with a set group of people on the composition of which she has no say. She can have better or worse relations with the people in that group, but who gets to join the group is out of her hands' (p. 128).

The question is if these enforced relationships can form the foundation of friendships. In this context, it is interesting that the protagonist despises in 'Curious George' (1.03) the whole concept of friendship because of its inherent arbitrariness: 'One desperate attempt after another to find something in common with someone else and then cling. Hey, you have ten fingers, I have ten fingers, let's be friends. We'll make rules and slogans. Then if we find someone with nine fingers, we can beat the crap out of them.' This critique of the arbitrariness of freely chosen human relationships outside the nuclear family makes way for the next argument: If social fabrics are not regulated by choice or destiny anymore, the alternative families of post-millennial television can be described instead as groups of specialists with each member enriching the group with a singular ability or a cohesive contribution. The group is structured by an established, but rather undemocratic hierarchy between the characters that is often questioned and twisted. In this sense, these expedient alliances are closer to the nuclear family than to friendships.

While the fields of expertise of the characters in *Firefly* or *Heroes* are clearly defined by their occupational title (e.g., captain or doctor) or their supernatural powers (e.g., can see the future or can fly) respectively, the case is more diffuse with *Dead Like Me* and *Pushing Daisies*. The organization of grim reapers is relatively undemocratic with Rube as their uncrowned king. Rube is the one who gets the list of people who are about to die and passes Post-its with the name and estimated time of death to his fellow reapers. Thus, the others assume that he has more knowledge, which justifies his superior status most of the time. Although the other reapers have exactly the same job, they are defined by different styles that contribute to their group as well as to the general philosophy of taking souls: George contributes with her sobriety and wit, Roxy with efficiency and anger, Mason with awkwardness and reveries, Daisy with her glamour and fragility. In *Pushing Daisies*, the characters are likewise assigned complementary attributes that help them to form a unity of individuals. In many aspects, the characters of *Dead Like Me* are adapted for *Pushing Daisies* with Ned as variation of Rube, Chuck mirroring George, Emerson mirroring Roxy, and Olive mirroring an unholy combination of Mason's clumsiness, Daisy's romanticism, and Betty's joie de vivre.

In the postmodern social networks that are portrayed in television series, every character needs a function to become a part of the community. It is insightful that even in *How I Met Your Mother*, a programme that immensely imitates *Friends* at first glance, this factor is clearly articulated. This alliance of people can only persist because they can be identified as 'specialists' in their fields. In 'The Ashtray' (8.17), Ted tells a story and Barney is desperate to become a part of it although it is beyond question that he did not participate in the original event. Barney's explanation for his attempt to rewrite history is telling: 'You have architecture, Marshall has the law, Lily has art, Robin has pleasing me sexually, you all have a passion that drives you! Well, I have a passion, it's taking life and turning it into a series of crazy stories. If you can do that without me, then I don't even know who I am anymore.' In contrast, the professions of Rachel, Joey, Monica, Chandler, Ross, and Phoebe were so irrelevant for the plot in terms of the interaction between the group of *Friends* that it is difficult to even recall them.

In closing, there is another important aspect of the portrayal of social networks in the post-millennial television series left to address that still operates in the same way as known from *Friends*. Similar to the Central Perk, the public spaces of Der Waffle Haus in *Dead Like Me* and The Pie Hole in *Pushing Daisies* are replacements of the domestic spaces—the kitchen and the living room—of earlier sitcoms and family dramas like *The Cosby Show*, *Who's the Boss?*, *Full House*, or *Growing Pains*. These spaces have been established as the essential sphere for family communication and bonding since *I Love Lucy.* Der Waffle Haus and The Pie Hole signify similarly to the Liberty Diner in *Queer as Folk*, MacLaren's in *How I Met Your Mother*, the Bada Bing in *The Sopranos*, The Cheesecake Factory and the cafeteria at Caltech in *The Big Bang Theory*, as well as the diner Merlotte's and the night club Fangtasia in *True Blood*, the dysfunctionality of the conventional family bond and the nonexistence of the domestic space as safe sanctuary. For both Chuck and

George, the domestic spaces have become dangerous and impassable; they are as lost to them as their biological families, displaced to the margins of the narration. The public space in television series is no longer a symbol for alienation and urban isolation (Parmiter, 2011); instead, the public space has moved to the heart of society and to the core of the narrations, providing a framework that is both professional and nurturing.

Nostalgic/Ironic/Postmodern

Television does not always offer a space where social anxieties are resolved as fairytale-like wish-fulfilment; but instead it can provide a space where the multifaceted nature of these anxieties can be discussed, explored, and dissected without the urge to present an ultimate resolution. The examples *Dead Like Me* and *Pushing Daisies* have illustrated that the family representations operate as a swan song for the nuclear family and establish expedient alliances as a blueprint for the postmodern family. But there is always a bittersweet atmosphere, suggesting that there is no proper compensation for the loss of the nuclear family. There is no offer of improvement, reparation, and redemption. The characters always experience a feeling of incompletion and inadequacy. On the one hand, they are driven by a nostalgic longing for 'the past as imagined, as idealized through memory and desire' (Hutcheon, 2000, p. 195). On the other hand, the protagonists have experienced the inadequacy of the nuclear family in their own biography and decline its conventions; they have become too dysfunctional themselves to ever be able to participate in a conventional relationship. Therefore, they are ground between the nostalgic longing for 'the way we never were' (Coontz, 2000) and its ironic deconstruction provoked by a profound distrust in traditional family arrangements. For example, the frequent flashbacks into an idealized past that are used as stylistic devices in both series combined with the dry voice-over-commentary that often contrasts the visual material point exactly to this friction.

The irresolvable contrariness between nostalgia and irony seems to be an inherent feature not only of the post-millennial television series but also of postmodernism itself. When Hutcheon (2000) shows how irony and nostalgia walk hand in hand in the postmodern, she uses postmodern architecture as example to illustrate her point but the representation of family in postmodern television series would be a similarly effective tool:

> [T]hink of the difference between contemporary postmodern architecture and contemporary revivalist (nostalgic) architecture; the postmodern architecture does indeed recall the past, but always with the kind of ironic

> double vision that acknowledges the final impossibility of indulging in nostalgia, even as it consciously evokes nostalgia's affective power. In the postmodern, in other words, (and here is the source of the tension) nostalgia itself gets both called up, exploited, and ironized. This is a complicated (and postmodernly paradoxical) move that is both an ironizing of nostalgia itself, of the very urge to look backward for authenticity, and, at the same moment, a sometimes shameless invoking of the visceral power that attends the fulfilment of that urge.

In the light of Hutcheon's analysis, think now of the difference between the family portrayals in contemporary postmodern television and in revivalist shows like *The Cosby Show* or the more recent *7th Heaven.* Like postmodern architecture, the postmodern television series recalls the past with an ironic double vision of the family that revivalist shows lack in their representation of relationships—no matter if they evoke an idealized family haven consisting of individuals related by blood or by friendship/choice. The postmodern television series constantly touches on the sensitive subject of how the individual in contemporary Western society is confronted with irreconcilable contradictions regarding the building up of relationships, the omission of the line of demarcation between work and private life, and the desperate but unaccomplishable desire for a 'final destination'.

References

Abrams, J. J. (Creator). (2004-2010). *Lost* [TV series]. ABC, Touchstone.

Aronsohn, L. & Lorre, C. (Creator). (2003). *Two and a Half Men* [TV]. CBS.

Arnaz, D. & Oppenheimer, J. (Producer). (1951-1957). *I Love Lucy* [TV series]. Desilu.

Baca Zinn, M. & Eitzen, D. S. (1987). *Diversity in Families*. New York: Harper & Row.

Ball, A. (Creator). (2001-2005). *Six Feet Under* [TV series]. HBO.

—. (Creator). (2008-). *True Blood* [TV series]. HBO.

Battis, J. (2005). *Blood Relations: Chosen Families in Buffy the Vampire Slayer and Angel*. Jefferson: McFarland.

Bays, C. & Thomas, C. (Creator). (2005-). *How I Met Your Mother* [TV series]. 20th Century Fox.

Branigan, E. (2007). Fokalisierung. *montage/av*, *16* (1), 71–82.

Burr, V. & Jarvis, C. (2007). Imagining the Family: Representations of Alternative Lifestyles in Buffy the Vampire Slayer. *Qualitative Social Work*, *6* (3), 263-280.

Busse, K. (2002). "Crossing the Final Taboo: Family, Sexuality, and Incest in Buffyverse Fan Fiction." In R. V. Wilcox & D. Lavery, *Fighting the Forces: What's at Stake in Buffy the Vampire Slayer* (pp. 207-217). Lanham: Rowman

& Littlefield.
Chase, D. (Creator). (1999-2007). *The Sopranos* [TV series]. HBO.
Cohan, M. & Hunter, B. (Creator). (1984-1992). *Who's the Boss?* [TV series]. Embassy, Cloumbia.
Coontz, S. (2000). *The Way We Never Were: American Families and the Nostalgia Trap*. New York: Basic Books.
—. (2005). *Marriage, A History: How Love Conquered Marriage*. New York: Viking/Penguin.
Cosby, B., Leeson, M. & Weinberger, E. (Creator). (1984-1992). *The Cosby Show* [TV series]. NBC.
Cowen, R. & Lipman, D. (Producer). (2000-2005). *Queer as Folk* [TV series]. Showtime.
Crane, D. & Kauffman, M. (Creator). (1994-2004). *Friends* [TV series]. Warner Bros.
Dauphin, M. (2011). "Pushing Daisies Away: Community Through Isolation." In A. Burger (Ed.), *The Television World of Pushing Daisies: Critical Essays on the Brian Fuller Series* (pp. 28-42). Jefferson: McFarland.
Euromonitor International. (2012, May 17). *Special Report: Rise in Single-Person Households Globally Impacts Consumer Spending Patterns*. Retrieved from http://etd.ohiolink.edu/send-pdf.cgi?bgsu1151353213
Franklin, J. (Creator). (1987-1995). *Full House* [TV]. Lorimar, Warner Bros.
Fuller, B. (Creator). (2003-2004). *Dead Like Me* [TV series]. MGM.
—. (Creator). (2007-2009). *Pushing Daisies* [TV series]. Warner Bros.
Gill, P. (2011)."It's a Destiny Thing—Enjoy It!": Free Will and Determinism in Bryan Fuller's Series. In A. Burger (Ed.), *The Television World of Pushing Daisies: Critical Essays on the Brian Fuller Series*. Jefferson: McFarland.
Goldberg, G. D. (Creator). (1982-1989). *Family Ties* [TV series]. Paramount.
Halberstam, J. (2005). *In a Queer Time and Place: Transgender Bodies, Subcultural Lives*. London, New York: New York UP.
Hampton, B. (Creator). (1996-2007). *7th Heaven* [TV series]. Spelling.
Herek, S. (Director).(2009).*DeadLike Me:Life After Death*[Feature film].MGM.
Hutcheon, L. (2000). "Irony, Nostalgia, and the Postmodern." In R. Vervliet & A. Estor (Ed.), *Methods for the Study of Literature as Cultural Memory* (pp. 189-207). Altanta: Rodopi.
Kring, T. (Creator). (2006-2010). *Herores* [TV series]. NBC.
Lawrence, B. (Creator). (2001-2010). *Scrubs* [TV series]. ABC, Touchstone.
Leavitt, R. & Moye, M. G. (Creator). (1987-1997). *Married... with Children* [TV series]. Sony.
Lorre, C. & Prady, B. (Creator). (2007-). *The Big Bang Theory* [TV series]. Warner Bros.
Marlens, N. (Creator). (1985-1992). *Growing Pains* [TV series]. Warner Bros.
Moore, K. H. (2006). *The Grim Reaper, Working Stiff: The Man, the Myth, the Everyday* (Master's Thesis). Retrieved from http://etd.ohiolink.edu/send-pdf.cgi?bgsu1151353213
Parmiter, T. K. (2011). "Sweet Talk in the Pie Hole: Language, Intimacy, and Public Space." In A. Burger (Ed.), *The Television World of Pushing Daisies:*

Critical Essays on the Brian Fuller Series (pp.155-173). Jefferson: McFarland.

Reo, D. (Creator). (1990-1995). *Blossom* [TV series]. Touchstone.

Sandell, J. (1998). I'll Be There For You: Friends and the Fantasy of Alternative Families. *American Studies*, *39* (2), 141–155.

Shapiro, A. J. (2011). *You Only Live Twice: The Representation of the Afterlife in Film* (Master's Thesis). Retrieved from http://scholarlyrepository.miami.edu/oa_theses/280

Stacey, J. (1996). *In the Name of the Family: Rethinking Family Values in the Postmodern Age*. Boston: Beacon Press.

Starr, D. (Creator). (1998-2004). *Sex and the City* [TV series]. HBO.

Stevens, K. (2010). Meet the Cullens: Family, Romance and Female Agency in Buffy the Vampire Slayer and Twilight. *Slayage*, *8* (1), 1-25.

Taylor, E. (1989). *Prime-Time Families: Television Culture in Postwar America*. Berkeley: University of California Press.

Whedon, J. (Creator). (1997-2003). *Buffy the Vampire Slayer* [TV series]. 20th Century Fox.

Whedon, J. (Creator). (2002-2003). *Firefly* [TV series]. 20th Century Fox.

Williams, M. (Creator). (1988-1997). *Roseanne* [TV series]. Winddancer, Paramount.

CHAPTER THIRTEEN

SATYAMEVA JAYATE AND THE STARDOM THAT LOOMS: EMOTIONAL STOCK MARKET

SONY JALARAJAN RAJ
AND ROHINI SREEKUMAR

Introduction

In the entertainment sector, visual media, particularly television, will be the most revolutionary medium altering the lifestyle of a family. In India, television started as a community experience build on social service to ignite changes in society. As Doordarshan, the government broadcasting channel, began to extend its programming, the Indian audience was redefining its entertainment experience with new visual pleasures on screen. Perhaps 1980–1990s was the golden period of Indian television, even though only two national channels were available, DD National and DD2 (DD Metro). As television channels' survival in this cluttered media market depends on the unique content they develop, most of the television outlets prioritize 'emotional appeal', whether in news or entertainment. No other medium appeals to the senses of an anonymous, heterogeneous, and scattered audience of a mass medium as television efficiently and effectively does. On the one hand, news channels concentrate on live disaster videos to engage the audience and evoke awe and despair, while on the other, entertainment television channels have articulated reality shows as their bridge to the audience, where large scope looms for emotional drama. The elements of uncertainty, curiosity, anxiety, and speculation form the ideal chemistry to manufacture this magic mix. The success of these reality televisions or any other reality chat shows (where the programme is shot in the presence of a studio audience) is based on emotions with which they play to reach the audience. This is manifested as SMS votes, emails to the editor, YouTube comments, Facebook pages, etc.

India has also recently witnessed an array of so called reality programmes inspired from (and adaptations of) western shows like *Big Brother* (*Big Boss* in India) and *Who Wants to be a Billionaire* (*Kuan Banega Crorepathi* in Hindi and several other regional versions of it) that have cluttered the entrainment media and revolutionized the entertainment desires of the Indian audience. When Bollywood and other regional film stars were competing to host these shows (Amitabh Bhachchan and Shah Rukh Khan) with high remuneration and perks, their television presence can be taken as just an extension of their pompous stardom in film, where they are prominent in every shot.

It is at this juncture that Aamir Khan, the perfectionist of the Indian film industry entered the television screen with his iconographic programme *Satyameva Jayate*, a social service reality programme. Here, the virtuous stardom of Aamir Khan (Bollywood actor) is employed for the wide marketing of social messages through a one-hour emotional discussion saga. In many cases, a celebrity's emotional imagery has been utilized for financial and political benefits and entertainment. However in *Satyameva Jayate*, this has been an ambitious step by Aamir's own production house for the formation of a public sphere for societal advancement. This study tries to locate this phenomenon of *Satyameva Jayate*, where two factors—stardom and emotional marketing—have been juxtaposed in the development of a new kind of public sphere. This chapter explores the success of the show based on the emotional appeal of a star on the mini-screen, how the programme acts as a new kind of public sphere.

What's *Satyamev Jayate* all about?

Satyameva Jayate (*Truth alone Prevails*) is the national motto of India, which is taken from the ancient Indian scripture, *Mundaka Upanish*. Though most of the news programmes claim to hold truth in their respective depictions, it was *Satyameva Jayate*, a socially committed programme that courageously adopted this name to make a difference in society. The show captured the public imagination, whereby the production team made sure that it reached maximum audience with maximum impact. For the first time in Indian television history, a show was telecasted simultaneously on a private channel, Star Network, and a public broadcasting channel, Doordarshan, to get maximum audience. During the initial period of establishing television in India, programmes were telecasted only for a few hours through tele-centres (like a club) having a television set where a community gathered to view the programmes. Following a similar model, in order to make it accessible to remote villages, *Satyameva Jayate* (henceforth *SJ*) was telecasted in community centers simultaneously with its national broadcasting.

Spanning only three months, each episode of *SJ* deals with one of the grave social issues in India which is sparingly debated in society. Unlike other shows, which keep either half or one hour as their duration (including advertisements), *SJ* made it mandatory that the running time (without advertisements) of the show must be one hour or more to delve deeply into each issue. Hence the show's duration comes to around one and a half hours with 20 minutes of advertisements. Taking valid witnesses and victims for each case, the discussions are supported by empirical and numerical evidence. The show also demanded audience participation whereby the audience could give their valid comments and suggestions through the programme's website or in the form of SMSs for the super question asked at the end of each episode. In order to decommercialize this attempt, each SMS would be converted to the value of Rs 1 (Ruppee), and would be donated to a trust or poor homes associated with the respective issues dealt with in each episode.

One of the major reasons for the international appreciation of the show is the presence of Aamir Khan, who was till then known for his proficiency as an actor in his internationally recognized films like *Lagaan* and *Mangal Pandey*. A major part of this article is on this new identity of Aamir Khan as a socially responsible citizen.

Empathetic Star on screen

The success of the show lies mostly in the way a big star has come down to the status of laymen, delving into the grievances of society, weeping and at times consoling the victims on screen with the reality of Indian society. The programme is a perfect example of a new form of marketing, where stardom is incorporated with emotion, perhaps more than what the Oprah Winfrey show promises. While many of the Bollywood stars have already marked their presence on the small-screen (for instance, Amitabh Bhachchan and Shah Rukh Khan), Aamir had the privilege of an international media attention, with magazines like Time making him one of the most popular personalities in the world. Why is it that Aamir became an all-time hit on the mini- screen, and why Aamir? There are two reasons for this unequaled social status that Aamir holds now. One lies in his stardom and star image, which is distinctive and unique from his contemporaries. Second, his television charisma, or 'Aamir effect' (*Aamir gash*) on screen.

It is often commented that stardom is a combination of a star's filmic and non-filmic narratives, where the star image is produced by a combination of texts like films, publicity programmes, criticisms, etc. (Dyer, 1998). Shah Rukh Khan would be a perfect example of such an international figure who conquered the global audience with his dancing skills, his stage presence, and his high energy. Aamir is an actor whose articulation of stardom is quite different from his contemporaries. Apart from his films, his presence could never be felt in any film awards or star shows, which are now the primary site of stardom after the classical mediums like newspaper and magazines (Mazumdar, 2012). Aamir has his own perception regarding attendance at these award nights, and has restricted his productions from applying for any awards (which is often the procedure to enter the list) after the success of his social satire, *Peepli Live*. As a response to this Aamir Khan opines,

If anyone wants to honour a genuine effort, they can take the initiative in doing so. To me, the biggest achievement is getting a good response from the audience ('Aamir Khan says', 2011, June 17).

While actors and celebrities vie to get their wax model to be placed in Madam Tussaud's, it was Aamir who turned down the offer saying that he has little interest in making a model of himself and he only wants his films to be watched by everyone, not his wax model ('Aamir Khan refuses', 2009, February 15). Aamir will be the least popular star in any Bollywood stage tours that go around different continents, hence pushing him out of the category of a stage entertainer. Hence the presence of Aamir is mostly felt only in his films and his film promotions. Moving on to the films, Aamir is often quoted as the 'perfectionist' of the Bollywood industry, though he claims that he merely leaves 'no stone unturned' (in his films) ('I don't Consider', 2013, April 1). After the establishment of his own production company, Aamir Khan Productions, he managed to be part (as an actor or producer) of only those films that can give some social messages or those dealing with social issues, or those films that strike the untouched realm of life and filmography. This is evident from his films starting from *Lagaan* (2001), *Mangal Pandey, the Rising* (2005), *Rang De Basanti* (2006), *Taare Zameen Par* (2007), *Peepli Live* (2010), and the list goes on. All of these films are either historical (a fictional treatment of the history of India) or social satires or edutainment films that gave some messages to the community. When moving on to television, most of the popular Bollywood stars like Amitabh Bhachchan, Shahrukh Khan, Akshay Kumar, etc. have entered the mini-screen to widen and exploit their stardom among television viewers. Neither of these Star presence in television was for any social activism, but for entertainment needs and prominence.

Being the descendant of the Indian freedom fighter and scholar, Abdul Kalam Azad, his zest for social service is not only evident in his films, but also in his social presence. His initial undertaking as a social activist can been seen in 2006, when he participated in the *Narmada Bachao Andolan* (Save Narmada Movement) headed by environmentalist Medha Patkar against raising the height of the Narmada Dam and displacing the tribal community from their natural environment and land. Though he never criticized the construction of the dam, but supported the rehabilitation of the tribal community, he had to face stiff resistance from the State of Gujarat, where his film *Fanaa* (2006) was not allowed in any cinemas across the State. When Anna Hazare, a social activist, started a Satyagraha model of protest for passing the anti-corruption bill in Parliament, Aamir stepped forward to support his protest.

These initial social undertakings and his clear stand on each social and political issue took Aamir into the limelight of a celebrity activist. It is quite unusual for a celebrity to initiate such a move when they are still active in their respective field. Though the emergence of actors as politicians is nothing new in the Indian political arena (Prasad, 1999; Srinivas, 2009), to remain as a social reformer and to maintain that social commitment even in his films is.

The real thrust to initiating a programme like *Satymeva Jayate* can be seen in the promotional and marketing activities of his film *3 Idiots* (2009), a film peering into the higher educational system in India and elsewhere. As a way of promoting films remote villages, Aamir Khan adopted a unique marketing method of making a disguised tour into villages and leaving the news after his departure from each village.

Considering all these cultural intermediaries of Bollywood, Aamir Khan has made his own definition of a stardom with his denial of extra-filmic glamorous extravaganzas. This was indeed a suitable period for him to take his productions to the level of a genuine diffusion into society with a media intervention. India has made use of film stars in public service campaigns, for example Aiswarya Rai for the eye donation programme and Amitabh Bhachan for the Pulse Polio Campaign, realizing the impact and the role that these film stars have in society. Hence when Aamir came to the small screen with his own production of discourse to society about itself, the audience didn't carry any disbelief, owing also to the star persona that he had managed till then.

When the show was first telecasted on television on 6 May 2013, the audience were anxiously awaiting their star's small-screen appearance. It could be right to say that the entire show has Aamir charisma, not only through his presence as an anchor or moderator, but through his own production company delving into the social issues. Even before the airing of the first episode, Aamir and his team made it clear that it was not a show, but a social campaign lasting 13 weeks, and he declared with his usual zeal that 'I don`t care about TRPs. It is important to reach out to the Indian audiences. If viewers want to see, they will see or else it is up to them' ('Meant for private', 2012).

This 90 minute talk show is an episodical analysis of India's forbidden realities like caste discrimination, child abuse, the dowry system, sex-selective abortion, disabled people, medical malpractice, etc., all of which hold immense possibilities for an emotional saga. A show that demands grave understanding and authority, while leaving a considerable space for its emotional articulation on screen, has been perfectly realized by Aamir, who had earlier made his debut into social activism. Moreover, if one considers this show as a reality programme, the primary success factor of any reality television or popular television is its play with emotions. Whether the contestants weeping at the devastating comment, or the Judges' furious pronouncements, or the farewell or elimination drama, emotions play a very crucial role in the marketing of a television show in the postmodern era.

The first episode of the show, started with a prologue by Aamir, proclaiming the actor's role and motivation for such an initiation. As Aamir walks through a deserted beach, the voice over speaks as his internal narration. With a filmic dramatic mood of a pensive character, the actor maintains expressions as the prologue continues. Through this he tried to take the audience to the genuiness of an actor like him on the social issues of India. This sets the mood and invites the audience to the experience of a reality show. He opines that he has played many characters, but he has also a life, the life of an ordinary man, not an actor; while at times, when he hear success stories regarding India, he feels proud, there are many other instances when we purposefully turn a blind eye towards certain social realities, because it hurts us. He began to speak to the audience.

> Sometimes I think about why I'm tormented by issues that have no direct relations with my life when my life is going perfectly. What makes the difference? But no. It makes a difference since I'm also part of this society… If Gandhi, Tilak, Bose, Nehru, Maulan Azad [India's freedom fighters] were alive, we couldn't have looked them in the eye, because when they fought for freedom, did they expect such an India of social evils? Hence I'm coming among you people, to hear you, to know more about you, to inform you... I want to bring all the truth out of its cocoon,

and give prominence to those issues that impact on all of us…

The parasocial relationship that audiences have with the stars on screen is intensified by the star's personal involvement and charisma, as he tries to shed the aura of stardom and comes to terms with social reality. As opposed to the other reality shows, where the presences of celebrities are part of commercial marketing, Aamir Khan in *SJ* unleashes an emotional journey to kick-start a new form of social service upon stardom. The prologue by Aamir with his sensuous emotional dialogues grabs the audience to the show as it pricks the political sentiments of a layman. Aamir, who is little connected to glamorous extravaganzas and extra-filmic events, took this opportunity to make a sincere effort to connect with the audience. During the release of its promo video, he made a confessing statement that he may not change the world, but he may realize the present social reality.

Through this show we understand the problem of the people, we are not here to make a change. I am no one to change anything... I feel understanding a problem and feeling it or holding one`s hand or hugging is also important. I may not have the solution, but at least I can hear and understand ('Aamir says he', 2012)

Even the choice of theme for the first episode, on Female Foeticide, is a strategic decision to evoke maximum emotion and sentiments in the initial connection with the audience. Holding on to the sentiments of a 'mother', Aamir has made this a weapon to outpour maximum sentiments, where he is sure that he can get a considerable amount of female, as well as male, audience from all demographics (from teenagers to elders). Starting with a sensuous intro of 'mother', as mother goddess from Hindu mythology and the concept of heaven under our mother's feet from Islam, the episode received an all-India 4.1 television rating across the channels (Varma, 2012, May 27).

Unlike other shows, which cement the anchor to a certain static physical position in the studio (either upright or in a chair), Aamir has made use of the studio floor as he freely moves towards the audience, and sits on the steps and among the audience, reducing the distance between a star and a layman. Sidelining all the screen presence of Aamir, the most judicious weapon ready in his hand as a versatile actor was celebrity tears, which, when amalgamated with weeping music and a heated audience, made magic on screen. As Langar (1981) stated, television personalities are distinguished for their authenticity, ordinariness, typicality, and familiarity, which is in contrast to the stardom that has a constructed image. Television fosters a close identification between persona and role—thus giving the impression that the TV personality is just being 'themselves'. There are times which clearly juxtapose the audience response (sometimes weeping, sometimes in a shock, sometimes curious) with the ongoing narration. Khan has used his versatility in acting to wipe away the tears, to cover his mouth out of anxiety or shake his head out of disbelief and stare at the narrator out of curiosity, making the mood dramatic as well as provoking. These acts efficiently create an 'intimacy' and 'authentic identities' (Murray, 2005, p. 129), a spontaneous outpouring similar to that of a layman.

The identity of a star comes at this juncture. The proliferation of the entertainment media has made these celebrities a major agent of social change (Singhal & Rogers, 1994). The effectiveness of Aamir as a perfect star model could be analyzed from Dyer's (1998) concept of stardom. As she opines, a star's offscreen personality is as important as his/her on-screen characters in shaping a perception about the star. These offscreen personalities are constructed through different cultural intermediaries and media texts like gossips, criticism, posters, reviews, award shows, stage performances etc. However for Aamir, his filmic cultural intermediaries are restricted, and concentrated more on his role as an activist and a perfectionist in an off-screen panel. Often the chief motive behind the struggle to achieve celebrity status is the desire to transcend the veridical self (Kellner 2003: 12). However for Aamir, his veridical as well as screen-self go parallel as his off-screen positioning as an activist, his filmic characters (or his films) with social messages, or his presence in a programme like *SJ* make his stardom quite uncomplicated. This, in turn, ties a ray of genuinity for the show, as he is alike in all his off-screen and onscreen presence. Apparently, it takes the audience's attention and persuasion to be a part of it. Here Aamir is not only an anchor, but this show is under the authority of the star, by whose production company it is made. Hence the success of the show echoes the success of the star.

Social activation through emotional marketing

While the initial step of audience persuasion has been bound up with the persona of the star, the initiation of a public sphere is a matter to be further pondered upon. The media has a tremendous role in the functioning of democracy, and hence been considered as the fourth pillar of Indian democracy. With the outset of new media like mobile phones and social networking sites, the formation of a public sphere and thereby, an informative and active citizenship has achieved more significance and practicability. According to Habermas, the public sphere is a social space where information and ideas can circulate and form an opinion, and initiate a move (1989). *SJ* can be correlated to this definition of public sphere. A major concern of any democracy is that every citizen, no matter their economic and demographic differences, has the right to get information and respond to any issues. However, even with the possibilities of the new media, many of the political and social discussions have never reached a major section of the people to date, mainly because of the digital divide and laggardness towards digitization (Fernandes, 2013). This was the major issue addressed by the show. Unlike other shows, which are marketed as a channel's pride endeavour, this show is telecasted by a private channel, Star Plus (which is available only for those who have a subscription to a cable network) and the national channel of India, Doordarshan (which is the national public service broadcaster reaching everywhere), and its regional telecasts in five languages . This was a significant step as it considered the lower section of the society who are denied a cable network because of their financial difficulties, thereby increasing the target audience. Moreover, in order to get maximum reach and importance, Aamir Khan started a weekly column in a national newspaper, *The Hindu*, which takes the shape of a self-reflection on each social issue dealt with that week. For those intellects in India, who claim not to watch television, this column takes them to the show.

Moreover, the show criticized mainly the Indian middle class entrepreneurs and business magnates in all fields who made the layman a scapegoat for various business deeds. This was indirectly connected to the inefficiency of the Indian government to tackle these money-mongers and anti-social institutions who have already been criticized for their corrupt governance.

One of the other significant steps taken by this show is the choice of time slot. Broadcasted on Sunday, the show was telecasted simultaneously on Doordarshan and Star plus at 11–12.30 pm, a momentous slot which used to be occupied by Indian historical dramas and epic narrations like *Ramayana* and *Mahabharatha*, and which claimed large number of family audience (Rajagopal, 2001) when public broadcasting channels like Doordarshan (DD1) and DD Metro (or DD2) were running the scene as the only television experience. The significance of the slot lies in its being not only a public holiday, but also a time when the family is together on a holiday morning.

As for the show itself, it was a democratic way of approaching an issue. Not only were the affected or victims of the respective issues brought to the studio, but the perpetrator or the offender was also given chance to state their comments and suggestions. That perfectly makes space for a public sphere. In one of the episodes against 'Intolerance to love' (in India love marriages are still considered a sin in many section of society), along with victims, members of the Khap Panchayat (the local administration having leaders of all castes in a particular council), who are determined to suppress all love marriages or inter-caste relationships, also took part. This was a momentous time, when the victim and the oppressor were on the same platform, leaving the audience to decide upon the future. Irrespective of language barriers, the research wing of the show collected data and stories from across India, and made their voice heard through this show.

To invite the audience's participation, the show opened a website, www.satyamevajayate.in, through which they can speak out and give their opinion regarding each social issues or their perception of the show. Through the 'Speak Up' section in the website, audiences can upload videos related to any social issues and share a story near their neighbourhood, ultimately making it a people's movement. Even in social networking sites, this show captured most of the spaces, being the most searched word on the first day of its telecast (Sheikh, 2012).

Conclusion

The success of the show is evident from the widespread appeals and moves by local and central administration towards curbing each social issues. For instance after the airing of the first episode on female foeticide, the Chief Minister of Rajastan (one of the States In India which has a high number of female foeticides), Ashok Gahlot, ordered elected members and NGOs to take stiff action against it. In another instance, the Khap Panchayat who supported honour killings for those who choose love marriage, found themselves at a threatening juncture, which made them appeal for the show to be taken off (Hussain, 2012).

As happens with any social event, the reception of the show was not devoid of criticism. A section of society claimed it was just a publicity gimmick by a star to get a huge ransom in return. Even the negative criticisms of the show turned out to be unexpected publicity for Aamir, as the sayings of Oscar Wilde goes—*the only thing worse than being talked about is not being talked about.* Though the programme has a quality and uniqueness of its own, the major factor that made it an unbeatable issue, conquering even the cover page of Time Magazine, is the host of the show, Aamir Khan. More than the show, the actor was singled out as one of the 100 most influential people in the world. Talk shows that deal with grave political and social issues are often anchored by bigwig journalists to inject legitimacy and gravity. Aamir's show's success is the best example of a celebrity's parasocial relationship with his audience. By correlating a film star's charisma and a television personality's ordinariness and familiarity, Aamir has positioned himself as a successful protagonist. When he comments, the image of an activist and perfectionist inspires people's support and activation from across the country.

Though his show could not make a tremendous change in society, people have become informed, and a wave of activation can be felt. His social commitment and popularity became so unbeatable that when Bloomberg Indian TV came forward with a debate show, *Outsider*, they had to chose a hard-hitting personality, Tim Sebastian, to host the show, who, though a world-renowned critic and television journalist, has little experience of a multi-religious and linguistic nation like India.

Most of the time, Aamir is at the epicenter of the show. His genuineness towards society is evident in the way he marketed his venture, through media platforms like newspaper and online media, with his personal note and signature. This doesn't take the shape of an advertisement, but of an editorial, as seen in his weekly column in *The Hindu*.

This paper has only given some initial observations into a star's manifestation of a social service touching on theories of stardom. However this leaves more opportunities for further research on a topic like social service through stardom making use of a celebrity's small screen persona and their filmic personae.

References

Aamir Khan refuses to get waxed at Madame Tussauds! (2009, 15 February). *Zee News*. Retrieved from http://zeenews.india.com/entertainment/celebrity/aamir-khan-refuses-to-get-waxed-at-madame-tussauds_23263.htm

Aamir Khan says no even to national awards! (2011, June 17). *India Today*. Retrieved from http://indiatoday.intoday.in/story/aamir-khan-national-awards-peepli-live-anusha-rizvi/1/141790.html

Aamir says he was jittery when he signed up for TV (n.p). Zee News. Retrieved from http://zeenews.india.com/entertainment/print.aspx?aid=109433

Anand, S. (2012, July 23). Silence eve jayate. *Outlook Magazine*. Retrieved from http://www.outlookindia.com/article. aspx?281646

Bennett, J & Holmes, S (2010): The 'place' of television in celebrity studies. *Celebrity Studies*, 1:1, 65–80

Ciecko, A (2001): Superhit hunk heroes for sale: Globalization and Bollywood's gender politics. *Asian Journal of Communication*, 11(2), 121-143. DOI: 10.1080/01292980109364807

Dyer, R (1998), *Stars*, 2nd edition, London: British Film Institute.

Ellis, J., 1982. *Visible fictions: cinema, television, video*. London and New York:

Routledge (original work published 1982).

Fernandes, K (2013, June 7). IT ministry working to bridge digital divide between urban and rural India: Deora. *Times of India*, 2013. Retrieved from http://articles.timesofindia.indiatimes.com/2013-06-07/goa/39814158_1_milind-deora-cell-tower-radiation-digital-divide

Habermas, J. (German) (1962) English Translation 1989). *The Structural Transformation of the Public Sphere: An Inquiry into a Category of Bourgeois Society*. (Thomas Burger, Trans.). Cambridge Massachusetts: The MIT Press,

Hussain, P. (2012, June 13). Khap Panchayat demands Aamir Khan's Satyamev Jayate to be taken off. *Merinews*. Retrieved from http://www.merinews.com/article/khap-panchayat-demands-aamir-khan.

I don't consider myself perfectionist: Aamir Khan (2013, April 1). *Times of India*. Retrieved from http://articles.timesofindia.indiatimes.com/2013-04-01/news-interviews/38163604_1_talaash-aamir-khan-productions-reema-kagti

Jain, K. (2007). *Gods in the bazaar: The economies of Indian calendar art*. Durham, NC: Duke University Press.

Langer, J (1981). Television's personality system. In T. O'Sullivan and Y. Jewkes (Eds), *The Media Studies Reader* (pp. 165–167). London: Arnold, 1997.

Mazumdar, R. (2012): Film stardom after liveness. *Continuum: Journal of Media & Cultural Studies*, 26:6, 833-844.

Meant for private, national channels, Aamir`s show to make history (2012). Zee News. Retrieved from http://zeenews.india.com/entertainment/print.aspx?aid=109414

Murray, S. (2005). *Hitch your antenna to the stars: early television and broadcast fame*. New York: Routledge.

Prasad, M (1999). Cine politics: On the political significance of cinema in south India. *Journal of the Moving Image* 1, 37–52.

Rajagopal, A. (2001). *Politics after television: Hindu nationalism and the reshaping of the public in India.* Cambridge, UK: Cambridge University Press.

Sheikh, M. (2012, May 11). 'Satyamev Jayate' gets its share of attention. *Livemint*. Retrieved from http://www.livemint.com/Consumer/tvHfSMtFZ66IZyaSClqdaM/8216Satyamev-Jayate8217-gets-its-share-of-attention.html

Srinivas, S. V. (2009). *Megastar Chiranjeevi and Telegu cinema after N.T. Rama Rao*. New Delhi: Oxford University Press.

CONTRIBUTORS

Barrette Pierre is Professor at École des médias, Faculté de communication, Université du Québec à Montréal and researcher at Centre de recherche interuniversitaire sur la littérature et la culture québécoises (CRILCQ). He wrote books and articles covering media theories, communication theories, history of television, Quebec television.

Belim Célia is a Teaching Assistent (*Assistente*) in the School of Social and Political Sciences, University of Lisbon (ISCSP-UL) (Courses: Political Agenda; Image Theory; Communication Theories; Press Techniques; Semiology; Social Psychology; Seminar; Digital Media among others) (since 2001), and a Ph.D candidate in communication studies with a thesis in progress on the Image of Portuguese front pages newspapers in the twenty-first century (since 2008). She has a master in political science (2008). Her master's thesis was about political propaganda after 9/11, focusing on analysis content of George W. Bush's and al-Qaeda leaders' speeches. Since January 2010, she has received a Portuguese Ministry of Higher Education Science and Technology doctorate grant. She has experience in image and media relations management (in the National Federation of Teaching and Research (FENEI / SINDEP), and in the Communication and Image Office (GCI) of ISCSP-UTL). And she had the best classification of communication sciences university degree (*licenciatura*). Belim has participated in conferences in her areas of interest, such as: the Midwest Political Science Association (MPSA), the International Political Science Association (IPSA), the European Consortium for Political Research (ECPR), the Spanish Association of Political Science and Administration (AECPA), the Latin American Association of Political Science (ALACIP), the I Congress Of Semiotics of Culture in the Complutense University of Madrid (GESC), the International Association for Media and Communication Research (IAMCR), Elections, Public Opinion, and Parties (EPOP) conference. She is author of articles and book's chapters. She was invited by Lexington Book to co-edit the book *Image of U.S. Presidential Administrations: The Cases of George W. Bush and Barack Obama*. Her major research interests are: analysis of image, political image and communication; political, public and media agendas; semiology; semiotics of culture;

filmic analysis; journalism and its relation with politics; and communication in its all expressions.

Branea Silvia is Associate Professor Ph.D at the Department of Cultural Anthropology and Communication, Faculty of Journalism and Communication Sciences, University of Bucharest and Associate Researcher at the Laboratory "Sociology of Communication and Public Space" (The Institute of Sociology of The Romanian Academy). A part from her teaching and research activities, she also collaborated with Radio Romania Cultural, the public radio broadcasting company. She was part of the team in research grants, both at national and international level. Main areas of research: communication and globalization, reception of TV series among youth, organizational sociology and analysis of online inter-ethnic forums. Her publications comprise author volumes (two single author books and two co-authored works), as well as studies published in Romanian and international journals (Romanian Journal of Journalism & Communication, Communication [Laval University], Journal of Community Positive Practices, Journal of Media Research, Reseaux).

Chiang Angie is a PhD candidate in the department of Communication and Culture at the University of Calgary. Her research and teaching interests include television consumption of contemporary complex narratives, convergence culture, third wave feminist media studies and dystopian films. Additionally, she has worked for various film, television, and interactive industries over the past decade in communications and marketing.

Colbran Marianne holds a Howard League/Oxford Post-Doctoral Fellowship. Before coming to the Centre she submitted a Ph. D in October 2011 entitled 'Watching the Detectives: A case study of production processes on 'The Bill' which explores how changing working processes, constraints of the medium and changing commercial imperatives all had an effect on representations of the police and policing on the show over its twenty-six year history. She studied for her PhD at the LSE. Before this, she worked as an actress and television scriptwriter, including seven years as a staff writer on The Bill.

Cuntz-Leng Vera (*1979) studied film and theatre science in Mainz, Marburg, and Vienna. Currently, she is a Ph.D. candidate in the Department of Film Studies at the Johannes Gutenberg-Universität Mainz in Germany. In her dissertation, Cuntz-Leng analyses the *Harry Potter* franchise on the intersection of genre theory, queer reading, and slash fandom. She has published on *Harry Potter*, Science Fiction and Fantasy cinema, Indian cinema, and Peter Weir.

Gueorguieva Valentina is assistant professor at the Department of Cultural Studies of the University of Sofia "St.Kliment Okrhidski". She holds a PhD in sociology from the University Laval (Quebec, Canada), and has specialized at the Center for the Studies of Social Movements at the *École des Hautes Études en Sciences Sociales* in Paris; at the Central European University in Budapest. Her research interests are in youth cultures and social movements, contemporary social and cultural theory, popular and media cultures.

Karadimitriou Achilleas, PhD candidate, Department of Communication and Media Studies, University of Athens. Achilleas Karadimitriou is a PhD candidate in the Department of Communication and Media Studies of Athens University. He is a Scholar of State Scholarships Foundation. In the same Department of Athens University he has completed a BA course. He also has a Master of Arts in Television Studies by the University of Wales - Aberystwyth. He is tutor in modules regarding television, online journalism and communication theories in public vocational training centers and colleges.

Marinescu Valentina is a Ph. D Reader at the Faculty of Sociology and Social Work – Bucharest University (Bucharest, Romania). She teaches undergraduate and graduate courses in media and society, and methods of researching mass communication. Her interests lie in media and communication studies, with a specific focus on East Asia. She has also published articles and book chapters on those subject matters. She was a fellow at the Academy of Korean Studies (Republic of Korea).

Mitu Bianca is a Lecturer at University of Bucharest. She has a PhD in Communication Science. Visiting Scholar at CAMRI, University of Westminster, Faculty of Arts, Media and Design, London in 2011. Her research interests lie in media and communication studies, with a specific focus on political communication and television studies. She has an outstanding international scientific activity and publications in International Journals from UK, Italy, Germany and USA. She is an active member of the most important international scientific associations (IAMCR, ECREA, UK Social Policy Association) and an active editorial board member of Journalism and Mass Communication Journal, USA and

International Journal Interpersona and International Journal *Management Dynamics in the Knowledge Economy*.

Petersson Åsa has a PhD in Child Studies at the department of Thematic Studies at Linköping University, Sweden. Her PhD project focused on the configuration of a child TV audience in policy documents and TV programmes for children.

Picard Yves is Permanent Professor at the Département de Communication, Cégep André-Laurendeau. He wrote articles covering media theories, communication theories, history of television, Quebec television.

Poulakidakos Stamatis, Specialized Research Personnel, Laboratory for Social Research on Mass Media, Department of Communication and Media Studies, University of Athens, stamatisp@media.uoa.gr. Stamatis Poulakidakos is a graduate of the Faculty of Communication and Media Studies University of Athens, where he is currently accomplishing his PhD. He also holds a Master's Degree in New Media, Information and Society by the London School of Economics. At the moment he works as Specialized Research and Teaching Personnel of the Laboratory for Social Research in Mass Media, Faculty of Media and Communication, University of Athens, specialized in media monitoring and analysis. At the same time he is responsible for EU funded internship programme of the undergraduate students of the Media Department. He has taken part in various research activities of the Laboratory for Social Research in Mass Media and has participated as speaker in Greek and international conferences. At the same time he works as a Professor, teaching Film and Media in the Public Sphere and Media Regulation Schemes at "Athens Metropolitan College".

Raj Jalarajan Sony is the Graduate Coordinator for MA Electronic Media and Assistant Professor of Communication Arts at the Institute for Communication, Entertainment and Media at St. Thomas University Florida, USA. Dr. Raj is a professional journalist turned academic who has worked in different demanding positions as reporter, special correspondent and producer in several news media channels like BBC, NDTV, Doordarshan, AIR, and Asianet News. Dr Raj has been a Faculty Member in Journalism, Mass Communication, and Media Studies at Monash University, Australia, Curtin University, Mahatma Gandhi University and University of Kerala. He is a three times winner of the Monash University PVC Award for excellence in teaching and learning. Dr Raj has been in

the editorial board of five major international research journals and he edits the Journal of Media Watch. Dr Raj was the recipient of Reuters Fellowship and is a Thomson Foundation (UK) Fellow in Television Studies with the Commonwealth Broadcasting Association Scholarship. Contact Details: Dr Sony Jalarajan Raj, Assistant Professor in Communication, Institute of Communication, Entertainment and Media, School of Leadership Studies, St. Thomas University, 16401 NW 37Avenue, Miami Gardens, Florida 33054, USA. Tel: 001 -305 474 6834 (Office) Cell: 001-786 – 204 -1031
Email: sony.raj@stu.edu, sonyjraj@gmail.com

Schluetz Daniela (PhD 2002) is lecturer at the Department of Journalism and Communication Research at the Hanover University of Music, Drama and Media. Her research focus is entertainment research (especially TV series), advertising communication as well as empirical research methods.

Schneider Beate (PhD 1973) is professor at the Department of Journalism and Communication Research at the Hanover University of Music, Drama and Media. Her research foci are national and international media systems, media politics and media production.

Sreekumar Rohini is pursuing her PhD from the School of Arts & Social Sciences at Monash University. She had her Master's Degree in Mass Communication and Journalism from Mahatma Gandhi University, India with a gold medal. Rohini is the recipient of National Merit Scholarship and Junior Research Fellowship from the University Grants Commission of India. Her research interest includes Journalism Practice, Mediated Public Sphere and Diaspora Studies. Contact Details: Ms. Rohini Sreekumar, Ph.D Research Scholar, School of Arts & Social Sciences, Monash University, Sunway Campus, Bandar Sunway, Malaysia 46150. Tel: 00601116326460.
Email: luk4rohini@gmail.com, rsre3@student.monash.edu

Taraghi Cherie, Ph. D. (Istanbul – Turkey). Her research focus on communication and media.

Veneti Anastasia, Teaching Fellow, University of Leicester *& Collaborating Researcher, Laboratory of Social Research in the Media, University of Athens,* anastasia_veneti@yahoo.com. Anastasia Veneti studied Politics and Economics at Goldsmiths' College, University of London and she acquired a Master degree in Communication from City

University, London. She holds a PhD from the Department of Media and Communication Studies, University of Athens. Her professional experience includes work as a research analyst in renowned Greek market research companies and as a communication consultant in various political campaigns. She is the author of one book and several articles on political communication and media representation issues. She has participated in many national and international conferences and she is a member of the Hellenic Political Science Association. She teaches at Akmi Metropolitan College and she collaborates as Researcher at the Laboratory for Social Research on Mass Media of the University of Athens.

INDEX